飞机人机与环境工程专业
汉英词汇手册

主编　党晓民　赵竞全　封文春

北京航空航天大学出版社

内 容 简 介

本词汇手册主要由飞机人机与环境工程专业涉及的传热学、工程热力学、空气动力学、飞机结构、环境控制、应急救生、供氧、水/废水、生活设施、内装饰、环控救生系统实验等专业理论与技术知识的汉英词汇组成。

本词汇手册主要供从事飞机人机与环境工程专业的工程设计人员使用，也可供学习相关专业的人员参考。

图书在版编目(CIP)数据

飞机人机与环境工程专业汉英词汇手册 / 党晓民，赵竞全，封文春主编. -- 北京 : 北京航空航天大学出版社，2016.12

ISBN 978-7-5124-2308-4

Ⅰ. ①飞… Ⅱ. ①党… ②赵… ③封… Ⅲ. ①飞机-人-机系统-词汇-手册-汉、英 Ⅳ. ①V221-62

中国版本图书馆 CIP 数据核字(2016)第 272687 号

飞机人机与环境工程专业汉英词汇手册

主编 党晓民 赵竞全 封文春

责任编辑 陈守平

*

北京航空航天大学出版社出版发行

北京市海淀区学院路 37 号(邮编 100191) http://www.buaapress.com.cn

发行部电话:(010)82317024 传真:(010)82328026

读者信箱: goodtextbook@126.com 邮购电话:(010)82316936

北京宏伟双华印刷有限公司印装 各地书店经销

*

开本:850×1 168 1/32 印张:13.5 字数:401 千字

2017 年 1 月第 1 版 2017 年 1 月第 1 次印刷 印数:1 000 册

ISBN 978-7-5124-2308-4 定价:59.00 元

前　　言

飞机人机与环境工程(简称“人机环”)专业是一门综合性很强的专业,不仅涉及传热学、工程热力学、空气动力学等基础学科理论和飞机结构的知识,而且涵盖了飞机环境控制、应急救生、供氧、水/废水、生活设施、内装饰、环控救生系统实验等众多专业性知识。随着国际技术交流的不断扩大,专业英语的作用越来越明显。因此,编写一本人机环专业汉英词汇手册不仅有助于该专业领域的工程设计人员查阅外文资料,提高专业技术能力,而且为与国外专家进行技术交流提供了极大的便利。

本词汇手册由党晓民、赵竞全和封文春主编。各部分内容分工如下:传热学和工程热力学:赵竞全、马兰;空气动力学:韩王超;飞机结构:王娟;环境控制系统:党晓民、李艳娜;控制和软件:张丹辉;防冰除雨系统:林丽;生活设施与内装饰:党亚、陈哲;应急救生系统和供氧系统:封文春、窦田天;水/废水系统:陈志东;环控救生系统实验:李俊伟。

全书由北京航空航天大学赵竞全教授和中航工业第一飞机设计研究院党晓民所长主审。贾博完成本词汇手册的统稿工作。

为方便工程设计人员使用本词汇手册,在各分支专业部分,特意作了简要的导语,同时编写了常用缩略语的英汉对照。

由于时间有限,书中漏误之处在所难免,欢迎读者批评指正,以便改进。

编　者

2016 年 10 月

目　　录

传热学

专业术语

A a

埃[①] angstrom Å

B b

半无限 semi-infinite
半无限大平板 semi-infinite plate
薄壁温度 film temperature
保温材料/隔热材料 thermal insulating material
保温效率 insulation efficiency
贝尔定律 Beer's law
贝克莱特数 Peclet number
比热容 specific heat capacity
毕渥准则 Biot modulus, Bi
壁面温度 wall-face temperature/wall temperature
边界层平均温度 boundary layer average temperature
边界层/附面层 boundary layer
边界层动量积分方程 momentum integral equation of the boundary layer
边界层厚度 boundary-layer thickness
边界层理论 boundary-layer theory
边界层能量方程 energy equation of the boundary layer
边界层能量积分方程 energy integral equation of the boundary layer
边界层质量连续方程 mass continuity equation of the boundary layer
边界条件 boundary condition

① 埃,长度的非法定计量单位,符号 Å。

变导热系数　variable thermal conductivity

表面粗糙度　surface roughness

表面对流换热系数　surface-convection heat-transfer coefficient

表面热流　surface heat flux

表面温度　surface temperature

波长　wavelength

波尔兹曼常数　Boltzmann constant

伯努利方程　Bernoulli equation

C　c

参考焓法　reference enthalpy method

残差　residual

层流　laminar flow/laminar

层流边界层　laminar boundary layer

层流边界层能量方程　energy equation of the laminar boundary layer

层流底层　laminar sublayer

层流关联式　laminar-flow correlation

层流区　laminar region

差分方程　difference equation

叉排　staggered tube rows

常微分方程　ordinary differential equation

翅片　fin

翅片面积　fin profile

翅片效率　fin efficiency

初始条件　initial condition

初始温度　initial temperature

传播　propagation

传热　heat transfer

传热过程　heat-transfer process

传热机理　heat-transfer mechanism

传热系数　coefficient of heat-transfer

传热效能　heat transfer effectiveness
粗糙管路　rough tube

D　d

达因[①]　dyne
大气　atmosphere air
大气温度　atmosphere temperature
大气压　atmosphere pressure
单位　unit
单原子气体　monatomic gas
氮气　nitrogen
当地声速　local sonic velocity/local velocity of sound
当量定律　law of equivalence
导热　thermal conductivity
道尔顿分压定律　Dalton's law of partial pressures
等熵压缩系数　isentropic compressibility
等温线　isothermal line/curve
等温压缩系数　isothermal compressibility
等效　equivalence
第二焓方程　the second enthalpy formula
第二类永动机　perpetual-motion machine of the second kind
第二熵方程　the second entropy formula
第三类永动机　perpetual-motion machine of the third kind
第三熵方程　the third entropy formula
第一类永动机　perpetual-motion machine of the first kind
第一熵方程　the first entropy formula
电磁力　magnetic force
电流　electric current
电能　electrical work/energy

---

① 达因:力的非法定计量单位。1 达因$=10^{-5}$牛。

定干度线　constant-quality line
定焓过程　constant-enthalpy process
定容/等容过程　constant-volume/isochoric process
定容比热①　constant-volume specific heat
定容加热循环/奥托循环　Otto cycle
定容吸热　constant-volume heat addition
定容增压比　pressure ratio at constant volume
定熵/绝热效率　isentropic/adiabatic efficiency
定熵-定容过程　isentropic and isochoric process
定熵-定压过程　isentropic and isobaric process
定熵过程　isentropic process
定熵函数　isentropic function
定熵流动　isentropic flow
定熵膨胀　isentropic expansion
定熵压缩　isentropic compression
定熵指数　isentropic exponent
定熵滞止参数/滞止参数　isentropic stagnation properties/stagnation properties
定熵滞止状态/滞止状态　isentropic stagnation state/stagnation state
定温/等温过程　constant-temperature/isothermal process
定温-定容过程　isothermal and isochoric process
定温-定压过程　isothermal and isobaric process
定性温度　qualitative temperature
定压/等压过程　constant-pressure/isobaric process
定压比热②　constant-pressure specific heat

---

① “定容比热”为旧称，在国标中其已改为“比定容热容”，对应英文为 specific heat capacity at constant volume。

② “定压比热”为旧称，在国标中其已改为“比定压热容”，对应英文为 specific heat capacity at constant pressure。

定压放热　constant-pressure heat rejection
定压吸热　constant-pressure heat addition
动参数　dynamic property
动焓　dynamic enthalpy
动能　kinetic energy
动温　dynamic temperature
动压　dynamic pressure
对比比容　reduced specific volume
对比态　corresponding state
对比态方程　equation of corresponding state
对比态原理　law of corresponding state/principal of corresponding state
对比温度　reduced temperature
对比压力　reduced pressure
对比状态参数　reduced property
对数平均温差　logarithmic-mean temperature difference
多变过程　polytropic process
多变指数　polytropic exponent/index
多孔塞　porous plug
多孔塞试验　porous-plug experiment
多孔塞形式　porous-plug type

E e

二阶差分/二阶导数　second-order differential
二阶偏微分　the second order partial differential
二维导热　two-dimensional heat conduction
二维导热方程　two-dimensional heat-conduction equation
二维稳态导热　two-dimensional steady-state conduction

F f

反射率　reflectivity
非流线体　bluff body

非牛顿流　non-Newtonian fluid
非稳态　unsteady state
非稳态传热　unsteady-state heat transfer
非稳态传热分析　unsteady-state heat transfer analysis
非稳态导热　unsteady heat conduction
非稳态导热方程　unsteady-heat-conduction equation
非稳态问题　unsteady-state problem
非圆形管道　noncircular channel
非圆形截面　noncircular cross section
分离变量法　separation-of-variable method
分离常数　separation constant
分离点　separation point
分析法　analytical method
分析解　analytical solution
封闭空间的自然对流　free/natural convection in enclosed spaces
风扇　fan
辐射/热辐射　radiation/heat thermal radiation
辐射四次方定律　radiation fourth power law
浮升力　buoyancy force
傅里叶定律　Fourier's law
傅里叶公式　Fourier equation
傅里叶准则　Fourier modulus, *Fo*
附面层/边界层分离　boundary-layer separation

G g

伽马射线　γ-rays
干涉仪　interferometer
高速流　high-speed flow
高温区域　high-temperature region
高斯误差函数　Gauss error function

格拉晓夫数/格拉晓夫准则　Grashof number, *Gr*[①]
各向异性材料　anisotropic material
管翅式　finned-tube arrangement
管道流动系统　tube-flow system
管内层流换热系数　the tube laminar heat-transfer coefficient
管束排列　tube-bundle arrangement
光滑管　smooth tube
光速　speed of light
光子气体　photon gas
过渡层　buffer layer
过渡区　transition region
过余温度　surplus temperature

H　h

海斯勒图/诺谟图　Heisler chart
焓　enthalpy
焓差　enthalpy difference
焓恢复系数　enthalpy recovery factor
黑度　emissivity
黑体　ideal radiator/blackbody
横排　transverse rows
宏观速度　macroscopic velocity
红外线　infrared
后向差分　backward difference
环境温度　environment temperature
环肋　circumferential fin
换热过程　heat-transfer process
灰体　gray body

---

① $Gr=\frac{l^3 g\alpha\Delta T}{\nu^2}$，其中$-\frac{\Delta\rho}{\rho}=\alpha\Delta T$。

恢复系数　recovery factor

混合对流　combined convection

混合对流分析　combined-convection analysis

J j

机理　mechanism

积分上限　the upper limit on the integral

基尔霍夫定律　Kirchhoff's law

极长管路　extremely long tube

集总参数法　lumped-capacity method

集总参数分析　lumped-capacity analysis

几何相似　geometrically similarity

几何形状　geometric shape

加热管束　heating tube bank

加热平板　heated plate

加热竖平板　heating vertical plat

加热水平管　heating horizontal tube

假设/假定　assumption

假设速度　fictitious/assumed velocity

假设温度　fictitious/assumed temperature

假设整个肋表面处于肋基温度下的散热量　heat which would be transferred if entire fin area were at base temperature

间隙/空隙空间　void/void space

间隙面积　void area

剪切力　shear stress

角系数　angle factor

接触面　contact surface

接触面积　contact area

接触热阻　thermal contact resistance

接触系数　contact coefficient

节点　node

近似法　approximate method
近似分析/相似分析　approximate analysis
近似解　approximate solution
进出口温度的算术平均值　arithmetic average of the inlet and outlet temperature
经典法　classical method
经验公式/经验关联式　empirical correlation/formula relation
经验摩擦因子　empirical friction factor
局部换热系数　local heat-transfer coefficient
局部努塞尔数　local Nusselt number
距离增量　distance increment
矩形肋　straight rectangular fin
绝对温度　absolute temperature
绝热壁面　adiabatic wall
绝热壁温　adiabatic wall temperature
绝热层　insulation
绝热层的临界直径　critical diameter of insulation
绝热层厚度　insulation thickness

K　k

柯尔伯恩比拟　Colburn analogy
可见光　visible light
空间坐标　space coordinate
控制容积　control volume

L　l

拉普拉斯传递方程　Laplace transform equation
拉普拉斯方程　Laplace equation
兰伯特定律　Lambert's law
雷诺比拟　Reynolds analogy

雷诺数 Reynolds number, *Re*[①]

肋/翅片 rib/fin

肋/翅片厚度 the thickness of the rib/fin, rib/fin thickness

肋/翅片深度 the depth of the rib/fin, rib/fin depth

肋/翅片效率 rib/fin efficiency

肋基面积 the base area of rib

肋片表面积 the total surface area of rib

肋片的纵剖面积 the profile area of the rib

肋片性能 rib performance

厘米 centimeter

离心力 centrifugal force

力 force

立体角 solid angle

力与动量平衡 force-and-momentum balance

连续方程 continuity equation/relation

连续温度分布 continuous temperature distribution

量纲 dimension

量纲分析 dimension analysis

量子 quantum

量子统计热力学 quantum-statistical thermodynamics

临界格拉晓夫数 critical Grashof number

临界雷诺数 critical Reynolds number

流场 flow field

流场相似 flow-field similarity

流动边界层 hydrodynamic boundary layer

流动边界层分析 fluid dynamics and boundary-layer analysis

流动边界层厚度 hydrodynamic-boundary-layer thickness

流体 fluid

---

① $Re=\frac{\rho vl}{\eta}=\frac{vl}{\nu}$。

流体动力学　fluid dynamics
流体分离过程　flow-separation process
流体黏性　viscosity of fluid
流体平均温度　the average temperature of fluid
流体温度　fluid temperature
流线　streamline
流向　flow direction

M　m

马赫数　Mach number
脉动速度　fluctuation velocity
脉动温度　fluctuation temperature
密度　density
密度变化　density variation
密度差　density difference
面积　area
摩擦系数　friction coefficient
摩擦阻力　frictional resistance

N　n

内表面　inside surface
内径　inside diameter
内热阻　internal/internal-conduction resistance
能量传递　energy transferring
能量方程　energy equation
能量平衡　energy balance
黏流　viscous flow
黏性/黏度　viscosity
黏性耗散　viscous dissipation
黏性耗散项　viscous-dissipation term
黏性加热　viscous heating

黏性加热项　viscous-heating term

黏性剪切力　viscous-shear force

黏性力　viscous force

黏滞力项　viscous force term

牛顿冷却定律/公式　Newton's law of cooling

牛顿流体　Newtonian fluid

牛顿运动第二定律　Newton's second law of motion

努塞尔数　Nusselt number, Nu①

暖气　radiator

𝒫 𝓅

抛物线速度分布　parabolic velocity profile

偏微分方程　partial differential equation

偏微分方程组　system of partial differential equations

频率　frequency

平板　flat plate

平板换热分析解　flat-plate analytical solution

平板问题　flat-plate problem

平均换热系数　average heat-transfer coefficient

平均努塞尔数　average Nusselt number

平均自然对流换热系数　average free-convection heat-transfer coefficient

普朗克常数　Planck constant, h②

普朗特数　Prandtl number, Pr③

---

① $Nu=\frac{Kl}{\lambda}$。

② $h=6.626\times10^{-34}$ J·S。

③ $Pr=\frac{\eta c_p}{\lambda}=\frac{\nu}{a}$。

Q q

气膜导热率　gas film conductance

气体常数　gas constant

气体动理学理论　kinetic theory of gas

气体辐射　gas radiation

前向差分　forward difference

前缘　the leading edge

强迫对流　forced convection

强迫对流换热系数　forced-convection heat-transfer coefficient

氢泡法　hydrogen bubble method

球坐标　spherical coordinate

R r

绕流　cross flow

热边界层　thermal boundary layer

热边界层厚度　thermal-boundary/layer thickness

热沉/冷源　heat sink

热辐射　thermal/heat transfer radiation

热交换器　heat exchanger

热扩散率/导温系数　thermal diffusivity

热量　quantity of heat

热流量　heat flow rate

热流体　hot liquid

热流线　heat flow line

热脉冲　heat pulse

热能　heat/thermal energy

热容　heat capacity

热射线　thermal radiation

热损失/热耗　heat loss

热线风速仪　hot-wire anemometer

热源　heat source

热阻　thermal resistance

容积膨胀系数　volume coefficient of expansion

入口段　entrance region

瑞利数/瑞利准则　Rayliegh number, *Ra*[①]

S s

三次多项式　cubic polynomial

三角形肋　triangular fin

三维导热方程　three-dimension heat-conduction equation

上表面　upper surface

声速　acoustic velocity/velocity of sound

湿周长　wetted perimeter

实际绝热壁温　actual adiabatic wall temperature

实际散热量　actual heat transferred

时间常数　time constant, τ

时间微分　time derivative

时间增量　time increment

时均速度　time averaged velocity/mean velocity

时均温度　time averaged temperature/mean temperature

石英纤维风速仪　quartz-fiber anemometer

势能　potential energy

试验方法　experimental method

试验数据　experimental data

竖直平板　vertical flat plate

数量级分析　order-of-magnitude analysis

数值传热学　numerical heat transfer

数值法　numerical method

数值分析　mathematical analysis

---

① $Ra=\frac{l^3\rho^2 c_p g\alpha\Delta T}{\eta\lambda}=\frac{l^3 g\alpha\Delta T}{\nu a}$。

数值解　numerical solution
数值解的稳定性和收敛性　stability and convergence of numerical solution
水力直径　hydraulic diameter
水平管　horizontal tube
顺排　in-line tube rows
瞬时速度　instantaneous velocity
瞬时温度　instantaneous temperature
瞬态边界条件　transient boundary condition
瞬态加热过程　transient heating process
瞬态冷却过程　transient cooling process
瞬态数值法　transient numerical method
瞬态温度响应　transient temperature response
瞬态自然对流加热　transient natural-convection heating
斯特潘-玻耳兹曼常数　Stefan-boltzmann constant
斯特潘-玻耳兹曼定律　Stefan-boltzmann law of thermal radiation
斯坦顿数　Stanton number, St①
松弛法　relaxation method/technique
速度分布　velocity profile/distribution
速度梯度　velocity gradient

$\mathcal{T}$　t

太阳常数　solar constant
特征尺寸　characteristic dimension
通用函数　arbitrary function
通用气体常数　universal gas constant
图表　graphical chart
图解法　graphical method

① $St=\frac{K}{\rho v c_p}$。

W w

外表面　outside surface
外径　outside diameter
外掠管道流　flow across cylinders
外掠管束流　flow across tube banks
外热阻　external resistance
微分　derivation
微分方程　differential equation
微观速度　microscope velocity
微米　micrometer
维恩位移定律　Wien's displacement law
温差　temperature difference
温度场　temperature field
温度场相似　temperature profile similarity
温度传播深度　temperature-wave penetration depth
温度分布　temperature profile/distribution
温度梯度　temperature gradient
温度微分　temperature differential
温度响应　temperature response
温降　temperature drop
温升　temperature rise
温室效应　greenhouse effect
温压　thermal potential difference
紊流　turbulent flow
紊流边界层　turbulent boundary layer
紊流充分发展区　fully turbulent region
紊流度　turbulence level
紊流区　turbulent region
稳态　steady state
稳态换热　steady-state heat transfer
稳态流动能量方程　steady-flow energy equation

稳态温度场 steady-state temperature yield
涡流导热 eddy thermal conductivity
涡流动量扩散率 eddy diffusivity for momentum
涡流黏度 eddy viscosity
污垢热阻 fouling resistance
无量纲 dimensionless
无量纲变量 dimensionless variable
无量纲参数 dimensionless parameter
无量纲分析 dimensionless analysis
无量纲温度分布 dimensionless temperature distribution
无量纲形式 dimensionless form
无量纲组合 dimensionless grouping
无内热源的稳态二维导热 two-dimension steady-state conduction without heat sources
无内热源的稳态一维导热 one-dimension steady-state conduction without heat sources
无黏流动 inviscid flow
无线电波 radio waves
无限长圆柱 infinite cylinder
无限大平板 infinite flat
物理机理 physical mechanism
物理系统 physical system

$\mathscr{X}$ x

X 射线 X rays
吸收率 absorptivity
下表面 lower surface
相似问题 similar problem
相似准则 similar principle
形状规则物体 regular-shaped solid
形状因子 shape factor
旋转机械 rotating machine

Y y

压降 pressure drop
压力梯度 pressure gradient
一维导热方程 one-dimensional heat-conduction equation
一维导热分析 one-dimensional heat-conduction analysis
一维瞬态导热 one-dimension transit-state conduction
一维稳态导热 one-dimension steady-state conduction
一维系统 one-dimensional system
迎风面积 frontal area
有内热源的稳态二维导热 two-dimension steady-state conduction with heat sources
有内热源的稳态一维导热 one-dimension steady-state conduction with heat sources
有限差分法 finite-difference method
有限差分方程 finite-difference equation
有限差分解 finite-difference solution
有限差分近似 finite-difference approximation
有效导热系数 effective thermal conductivity
圆形截面 circular cross section
圆柱坐标 cylindrical coordinate
运动方程 equation of motion/motion equations

Z z

真空 perfect vacuum
真实壁温 actual wall temperature
正交方程 orthogonal function
正弦分布 sine-wave distribution
正弦温度分布 sine-wave temperature distribution
直径 diameter
指数 exponent
质扩散率 diffusivity

质量传递　mass transfer
质量连续方程　mass continuity equation
质量流量　mass flow rate
滞止参数　stagnation property
滞止焓　stagnation enthalpy
滞止温度　stagnation temperature
滞止压力　stagnation pressure
中等温差　moderate temperature difference
重量力　weight force
轴向　axial direction
轴向系统　axial system
主流温度　mainstream temperature
转捩　transition
准则方程　criteria equation
紫外线　ultra violent
自然对流　free/natural convection
自然对流边界层　free-convection boundary layer
自然对流换热　free-convection heat transfer
自然对流换热系数　free-convection heat-transfer coefficient
自然对流与强制对流并存的混合对流　combined free and forced convection
自由流/主流　free flow
自由流静温　static free-stream temperature
总传热系数　overall heat-transfer coefficient
总焓　total enthalpy
总热阻　overall thermal resistance
纵掠平板　flow along a flat plate
阻力　drag force
阻力系数　drag coefficient

工程热力学

专业术语

A a

阿伏伽德罗定律　Avogadro's law

B b

百分比湿度　percentage humidity
磅力[①]　pound-force
饱和　saturation
饱和空气　saturated air
饱和曲线　saturation line
饱和温度　saturation temperature
饱和压力　saturation pressure
饱和液体　saturated liquid
饱和液体线　saturated liquid line
饱和蒸汽　saturated vapor
饱和蒸汽压　saturation pressure
饱和蒸汽线　saturated vapor line
爆破压力　burst pressure
爆炸压力　explosion pressure
背/反压　back pressure
泵　pump
比参数　specific property
比定容热容　specific heat capacity at constant volume
比潜热　latent heat
比热容　specific heat capacity

---

① 磅力为英美制质量或重量单位，符号 lb。1 磅合 0.4536 千克。

比热比　ratio of the specific heat capacity
比热容关系式　specific heat relation
比容　specific volume
比重[①]　specific weight
闭口系　closed system/control mass
闭口系能量方程　energy equation for closed system
闭式系统　closed system
边界　boundary
变比热　variable specific heat
变比热容　variable specific heat capacity
标准大气压　standard atmosphere pressure
标准流量　standard flow
标准生成焓　standard enthalpy of formation
标准生成自由焓　standard free enthalpy of formation
标准状态　standard condition
表面力　surface force
表面张力　surface tension
表压　gage pressure
冰点　ice point
冰箱　refrigerator
伯努利方程　Bernoulli equation
伯斯洛特方程　Bethelot equation
不可恢复总压降　total nonrecoverable pressure drop
不可逆过程　irreversible process
不可逆膨胀　irreversible expansion
不可逆性　irreversibility
不可用能　unavailable energy
不可能不付代价地把热量从一个低温物体传给另一个高温物体

---

① 比重为旧称，在国标中已弃用，已改为“相对体积质量（relative volumic mass）”或“相对密度（relative density）”。

It is impossible for a self-acting machine, unaided by an external agency, to convey heat from one body to another at a higher temperature

不可能建造这样一个循环工作的机器,这机器除了从单一热源吸热和举起重物(作功)之外,而不引起其他的变化

No cycle process is possible whose sole result is the flow of heat from a single heat reservoir and the performance of an equivalent amount of work on a work reservoir

C c

参考点　reference point

参考状态　reference state

查理定律　Charle's law

长度　length

超临界工作状态　supercritical state

超临界郎肯循环　supercritical Rankine cycle

超声速流动　supersonic flow

冲压空气温升　ram air temperature rise

初始状态　initial/original state

除湿　dehumidification

除湿器　dehumidifier

D d

达因　dyne

大卡/千卡　great/large calorie/kilocalorie/kilogram calorie

大气温度　atmosphere temperature

单位　unit

单原子气体　monatomic gas

氮气　nitrogen

当地声速　local sonic velocity/local velocity of sound

当量定律　law of equivalence

导热　thermal conductivity
道尔顿分压定律　Dalton's law of partial pressures
等熵压缩系数　isentropic compressibility
等温线　isothermal line/curve
等温压缩系数　isothermal compressibility
等效　equivalence
第二焓方程　the second enthalpy formula
第二类永动机　perpetual-motion machine of the second kind
第二熵方程　the second entropy formula
第三类永动机　perpetual-motion machine of the third kind
第三熵方程　the third entropy formula
第一类永动机　perpetual-motion machine of the first kind
第一熵方程　the first entropy formula
电磁力　magnetic force
电流　electric current
电能　electrical work/energy
定干度线　constant-quality line
定焓过程　constant-enthalpy process
定容/等容过程　constant-volume/isochoric process
定容比热[①]　constant-volume specific heat
定容加热循环/奥托循环　Otto cycle
定容吸热　constant-volume heat addition
定容增压比　pressure ratio at constant volume
定熵/绝热效率　isentropic/adiabatic efficiency
定熵-定容过程　isentropic and isochoric process
定熵-定压过程　isentropic and isobaric process
定熵过程　isentropic process
定熵函数　isentropic function

---

① “定容比热”为旧称，在国标中其已改为“比定容热容”，对应英文为 specific heat capacity at constant volume。

定熵流动　isentropic flow
定熵膨胀　isentropic expansion
定熵压缩　isentropic compression
定熵指数　isentropic exponent
定温/等温过程　constant-temperature/isothermal process
定温-定容过程　isothermal and isochoric process
定温-定压过程　isothermal and isobaric process
定性温度　qualitative temperature
定压/等压过程　constant-pressure/isobaric process
定压比热①　constant-pressure specific heat
定压放热　constant-pressure heat rejection
定压吸热　constant-pressure heat addition
动参数　dynamic property

E e

尔格②　erg
二次能源　secondary energy source/secondary energy
二次再热　doublereheat/second reheat

F f

发光强度　luminous intensity
范德瓦尔斯常数　the van der waals coefficient
范德瓦尔斯对比态方程　the Van der Waals equation of corresponding state
范德瓦尔斯方程　the Van der Waals equation/equation of state
范德瓦尔斯力　the Van der Waals force
范德瓦尔斯气体　the Van der Waals gas

---

① “定压比热”为旧称，在国标中其已改为“比定压热容”，对应英文为 specific heat capacity at constant pressure。

② 功和能量的非法定计量单位，符号 erg。1 erg=1d yne · cm，1J $=10^7$ erg。

非孤立系　nonisolated system
非均匀系　nonuniform system
非平衡态　nonequilibrium state
沸点　steam point
分容积　partial volume
分熵　partial entropy
分压力　partial pressure
分子　molecular
分子动力学能　molecular kinetic energy
分子间引力　intermolecular force/interaction
分子势能　molecular potential energy
分子数　the number of molecular
分子运动论　kinetic molecular theory
负效应　negative effect
负压　negative pressure
负压区　negative pressure area

G g

干饱和蒸汽　dry-saturated vapor
干空气　dry air
干球　dry bulb
干球温度　dry-bulb temperature
干球温度计　dry-bulb thermometer
干湿球温度计/干湿计　psychrometer
干式冷却　dry cooling
干蒸汽　dry vapor
高度　height
功　work
工程热力学　engineering thermodynamics
工作压力　operating pressure
工作状态　operating state

功率　power

功重比　power to weight ratio

孤立体系/孤立系　isolated system

孤立体系熵增原理　increase of entropy principal

孤立体系㶲减原理　decrease of exergy principal

固相/态　solid phase/state

过程　process

过程方程　process relation

过热度　degree of superheat

过热器　superheater

过热液体　super saturated liquid/superheated liquid

过热蒸气/汽　super saturated vapor/super-heated vapor

H　h

亥姆霍兹函数　Helmholtz function

含湿量/比湿度　humidity ratio/specific humidity

焓-熵图/莫里尔图　H-S diagram/Mollier diagram

焓湿图/湿度图/温湿图　psychrometric chart

耗功系统　work-absorbing system

耗散效应　dissipative effect

喉部　throat

华氏温标/温度　Fahrenheit scale/temperature

华氏温度计　Fahrenheit temperature scale/thermometer

化学能　chemical energy

环境/外界　surrounding/environment

换热器　heat exchanger

回热　regeneration/recuperation

J　j

机械能　mechanical work

积分节流效应　integral throttling effect

基准　datum

基准温度　datum temperature

吉布斯定律　Gibbs law of partial entropies

吉布斯函数　Gibbs function

加热器　heater

加湿　humidification

加湿器　humidifier

焦耳/焦[①]　joule

焦-汤系数/焦耳-汤姆逊系数　Joule-Thomson coefficient

焦-汤效应　Joule-Thomson effect

校准流量　calibrated flow

节流过程　throttling process

节流式热量仪　throttling calorimeter

节流装置　throttling device

进气阀　intake valve

经典热力学　classical thermodynamics

净功　net work

静态/平衡状态/平衡态　static state/equilibrium state

静温　static temperature

静压　static pressure

静压差　static pressure differential

距离　distance

绝对湿度　absolute humidity

绝热　adiabatic

绝热饱和过程　adiabatic saturated process

绝热饱和温度　adiabatic saturated temperature

绝热过程　adiabatic process

绝热混合过程　adiabatic mixing process

绝热节流过程　adiabatic throttling process

---

① 焦耳表功、能量和热的单位，符号 J。1 J=1 N·m。

绝热流动　adiabatic flow

绝热膨胀过程　adiabatic expansion process

绝热系统/绝热系　adiabatic system

绝热效率/定熵效率　adiabatic/isentropic efficiency

绝压/绝对压力　absolute pressure

均匀系统/均匀系　uniform system

均匀状态定态流动过程　uniform-flow process

K k

卡/卡路里①　calorie/cal

卡诺定理　Carnot principal

卡诺热泵　Carnot heat pump

卡诺热机/卡诺机　Carnot heat engine/Carnot cycle engine

卡诺效率　Carnot efficiency

卡诺循环　Carnot cycle

卡诺制冷机　Carnot refrigerator

开尔文-普朗克说法　Kelvin-Planck statement

开口系/控制体积　open system/control volume

开口系一般瞬态能量方程　transient energy equation for open system

开式系统　open system/control-volume system

可变基准温度　various datum temperature

可逆　reversibility

可逆功　reversible work

可逆过程　reversible process

可逆绝热过程　reversible and adiabatic process

可逆系统　reversible system

可逆循环　reversible cycle

可压缩液体　compressed liquid

可用能　available energy

---

① 卡:卡为卡路里的简称,是热量的非法定计量单位,符号 cal。使 1 克纯水的温度升高 1 ℃所需要的热量就是 1 卡。

克拉贝龙方程　Clapeyron equation

克劳修斯不等式　Clausius inequality

克劳修斯说法　Clausius statement of the second law of thermodynamics

克劳修斯积分不等式　Clausius inequality/inequality of Clausius

空气标准循环　air standard cycle

空气压缩制冷循环　air compression refrigeration cycle

控制容积/开口系　control volume/open system, C. V.

控制质量/闭口系　control mass/closed system, C. M.

跨声速流动　transonic flow

扩压管　diffuser pipe

$\mathscr{L}$　*l*

拉伐尔喷管　Laval nozzal/converging-diverging nozzal

兰式温标　Rankine scale/Rankine temperature scale

朗肯循环　Rankine cycle

冷吨　ice ton/commercial ton/ton of refrigeration

冷量炕　anergy of coldness

冷量㶲　exergy of coldness

冷凝/凝结或液化　condensation/liquefaction

冷凝过程　condensation process

冷凝器　condenser

冷凝潜热/冷凝焓　latent heat/enthalpy of condensation

冷凝温度　condensing temperature

冷却塔　cooling tower

冷却效率　cooling efficiency

冷效应　cold effect

厘米　centimeter, cm

理想气体/完全气体　ideal gas/perfect gas

理想气体混合物　mixtures of ideal gas/ideal-gas mixture

理想气体温标　ideal-gas temperature scale

理想气体状态方程　ideal-gas equation of state

力　force
连续方程　continuity equation
量纲一的量　quantity of dimension one
临界比容　critical specific volume
临界比容比　critical specific volume ratio
临界参数　critical property
临界点　critical point
临界截面　critical section
临界流动　critical flow/sonic flow
临界温比　critical temperature ratio
临界温度　critical temperature
临界压比　critical pressure ratio
临界压力　critical pressure
临界声速　critical velocity of sound
临界状态　critical state
零效应　zero effect
流动能　flow work
流量　flow/flow rate
露点　dew point
露点计　dew meter
露点温度　dew point temperature

M　m

马赫数　Mach number
马力　horse power
麦克斯韦等面积法则　Maxwell's equal-area rule
麦克斯韦关系式　Maxwell relations
麦克斯韦第一关系式　first Maxwell relationship
麦克斯韦第二关系式　second Maxwell relationship
麦克斯韦第三关系式　third Maxwell relationship
麦克斯韦第四关系式　fourth Maxwell relationship

摩尔　mole
摩尔参数　molar property
摩尔成分　mole component
摩尔分数　mole fraction
摩尔气体常数/通用气体常数　universal gas constant
摩尔热容　molar heat capacity
摩尔数　the number of moles
摩尔质量　molar mass

N　n

耐压压力/验证压力　proof pressure
内部不可逆　internally irreversibility
内部可逆　internally reversibility
内部损失　internal loss
内能　internal energy
能量传递　energy transferring
能量方程式　energy equation
能量利用因数　utilization factor
能量平衡　energy balance
能量平衡分析/能平衡　analysis of energy balance
能量转化与守恒定律　law of energy conservation and conversion
能量转换　energy transformation
逆布莱顿循环　reversed Brayton cycle
逆卡诺机　reversed Carnot cycle engine
逆卡诺循环　reversed Carnot cycle
逆热机　reversed heat engine
逆向循环/制冷循环/热泵循环　reversed cycle/refrigeration cycle/heat pump cycle
凝固　solidification/freeze
凝固潜热/凝固焓　latent heat/enthalpy of solidification
凝固线　freezing line

凝华潜热/凝华焓　latent heat/enthalpy of desublimation
凝华曲线　desublimation line/curve
凝露　condensation
牛顿①　newton
牛顿米②　newton-meter

O o

欧拉法　Euler approach

P p

帕斯卡③　Pascal
膨胀　expansion
膨胀功　expansion work
平衡　equilibrium
平衡态　equilibrium state
平衡态热力学　equilibrium thermodynamics
平衡状态/平衡态/静态　equilibrium state/static state
平均/折合摩尔质量　average/apparent molar mass
平均/折合气体常数　average/apparent gas constant
平均比热容　average specific heat capacity
平均分子量　average molecular weight
平均误差　average error

Q q

气态　gaseous state
气体动力循环　gas power cycle
气体动理学理论　kinetic theory of gas

---

① 力的单位，符号 N。
② 符号 N·m。
③ 压强的单位，符号 Pa。简称帕。1 Pa=1 N/m^2。

气体热力学参数　thermodynamics property of gas
气相区域/过热蒸汽区　superheated vapor region
汽化　vaporization
汽化潜热/汽化焓　latent heat/enthalpy of vaporization
汽液两相共存区/湿蒸气区　liquid-vapor region/wet region
千瓦[1]　kilowatt
潜能　latent energy
潜热　latent heat
压力计/气压计　manometer

$\mathscr{R}$ r

热　heat
热泵　heat pump
热泵循环　heat pump cycle
热沉　heat sink
热导体　heat conductor
热电比　heat-to-power ratio
热功当量　mechanical equivalent of heat
热化学　thermochemical
热机　heat engine
热机循环/动力循环/正向循环　heat engine cycle/power cycle/forward cycle
热力学　thermodynamics
热力学参数　thermodynamics parameter
热力学第零定律　the zeroth law of thermodynamics
热力学第一定律　the first law of thermodynamics
热力学第二定律　the second law of thermodynamics
热力学第三定律　the third law of thermodynamics
热力学概率　thermodynamics probability

---

① 功率的单位，符号 kW。1 W=1 J/s。

热力学基本关系　basic thermodynamics relation

热力学绝对温标/热力学温标/开尔文温标　thermodynamics absolute temperature scale

热力学能/内能/内部储存能　thermodynamic energy

热力学普遍关系式　thermodynamics property relation

热力学摄氏温标　thermodynamics Celsius scale

热力学湿球温度　thermodynamics wet-bulb temperature

热力学势　potential of thermodynamics

热力学特性　thermodynamics property

热力学温标　thermodynamics scale of temperature

热力学温度　thermodynamic temperature

热力学系统　thermodynamics system

热力学性质相似/热力学相似　thermodynamics similarity

热力循环　thermodynamics cycle

热量　quantity of heat

热量分析法　heat analysis

热量炕　anergy of heat

热量烟　exergy of heat

热量仪　calorimeter

热流　heat flow

热能　thermal energy

热平衡　thermal equilibrium

热容量/热容　heat capacity

热效率　thermal efficiency

热效应　heat effect

热源　heat source

容积成分　volume component

容积式压气机　displacement compressor

容积效率　volumetric efficiency

熔解　melting/fusion

熔解潜热/熔解焓　latent heat of fusion/enthalpy of fusion

熔解曲线　melting/fusion line
熔解温度　melting temperature

S s

三相　the three phase
三相点　triple point
三相线　triple line
熵　entropy
熵产　entropy generation
熵方程　entropy equation
熵流　entropy flux/entropy transfer
熵判据　entropy criterion
熵增　entropy production
上转变温度　top inversion temperature
设计温度　design temperature
摄氏百分温标　Celsius scale
摄氏度　degree Celsius
升华　sublimation
升华潜热/升华焓　latent heat/enthalpy of sublimation
升华曲线　sublimation line
生成焓　enthalpy of formation
生成自由焓　free enthalpy of formation
湿饱和蒸气/湿蒸气　wet-saturated vapor/wet vapor
湿度　humidity
湿度图/焓湿图/温湿图　psychrometric chart
湿空气　moist air
湿球　wet bulb
湿球温度　wet-bulb temperature
湿球温度计　wet-bulb thermometer
湿球温降　wet-bulb depression
湿蒸气干度　quality of wet vapour

实际过程　actual process
实际气体　actual gas
势能　potential energy
水泵　water pump
水银/汞　mercury
水银柱　a mercury column
水蒸气　water vapour/vapor
水蒸气压　water vapour pressure
速度　velocity/speed
算术平均温差　arithmetic-mean temperature difference
算术平均温度　arithmetic-mean temperature

T　t

特征函数　eigenfunction
体积　volume
体积分数　volume fraction
体积分析法　volumetric analysis
体积功　volume work
体积流量　volume flow rate
体积热容　volumetric heat capacity
通风机　fanner ventilator
通用气体常数　universal gas constant, R
统计热力学　statistical thermodynamics
透平/涡轮　turbine/turbomachinery
推动功　propulsive work
推进效率　propulsive efficiency
推重比　thrust to weight ratio

U　u

U 型管　U-tube
U 型管气压计　the U-tube type of manometer

W w

瓦　watt[1]
外部能量　extrinsic energy
完全气体混合物　ideal gaseous mixture
微分节流效应　differential throttling effect
微观热力学　microscopic thermodynamics
维里状态方程　virial equation of state
温标　temperature scale
温度　temperature
温度恢复系数　temperature recovery coefficient
温度计　thermometer
温度效应　temperature effect
温降　temperature drop
温-熵图　T-S chart
温升　temperature rise
紊流　turbulent
稳态　steady-state condition
无量纲量　dimensionless quality
物理标准状态　standard reference state of physics

X x

下转变温度　bottom inversion temperature
限流环　restrictor ring
相对湿度　relative humidity
悬式温度计　sling thermometer
悬式干湿计　sling psychrometric
循环　cycle

① 功率的单位，符号 W。1 W=1 J/s。

Y y

压比/增压比/降压比/扩压比　pressure ratio
压焓图　p－h diagram/p－h chart
压降　pressure drop
压力　pressure
压力功　reversible steady-flow work
压力管　manometer tube
压力恢复　pressure recovery
压力恢复因数　pressure recovery factor
压力计　manometer/pressure gauge
压气机　compressor
压气机绝热效率/压气机效率　isentropic efficiency of a compressor
压容图/示功图　P－V diagram/work diagram/P－V chart
压缩　compression
压缩比　compression ratio
压缩功　compression work
压缩因子　compression coefficient
亚临界工作状态　subcritical state
亚稳态　metastable condition
亚声速流动　subsonic flow
液化/冷凝或凝结　condensation/liquefaction
液态　liquid state
液体　liquid
液相　liquid phase
液相区域　liquid region
液柱　a column of a liquid
液柱高度　the column height
英尺①　feet

---

① 英尺为英美制长度单位，1 英尺合 0.3048 米。

英热量单位　british thermal unit,Btu

㶲/最大可用功　exergy/maximum available work

㶲平衡分析法/㶲分析　analysis of exergy balance

㶲损失　exergy destruction

㶲效率　exergy efficiency

永动机　perpetual-motion machines

有效功　effective work

有效容积　effective volume

有效温度　effective temperature

有用功　available work/useful work

原子　atom

原子能　nuclear energy

Z z

再热　reheat

再热器/回热器　reheater

再热温度　reheat temperature

再热循环　reheat cycle

再热压力　reheat pressure

增压　supercharge

增压器　supercharger

兆瓦①　megawatt

真空　vacuum

真空度　the magnitude of the vacuum

真实比热容　real specific heat capacity

蒸发冷却过程/绝热加湿过程　evaporative cooling/adiabatic humidification

---

① 功率单位,符号为 MW。

蒸发器　evaporator

蒸发温度　evaporative temperature

蒸发压力　evaporative pressure

蒸气表　steam table

蒸气管　steam pipe

蒸气区域/气相区域　vapor area

蒸气图标　vapor chart

蒸气压力　vapor pressure

蒸汽/蒸气　vapour

蒸汽动力循环　vapour power cycle

蒸汽机　steam engine

蒸汽卡诺循环　the Carnot vapor cycle

正效应　positive effect

正压区　positive pressure area

制冷剂　coolant

质量　mass

质量参数　mass property

质量成分　mass component

质量定容热容　massic heat capacity at constant volume

质量定压热容　massic heat capacity at constant pressure

质量力　body force

质量流量　mass flow rate

质量守恒方程　equation of mass conservation

重度　specific weight

重力　gravity/the force of gravity

重力加速度　the acceleration of gravity

重量　weight

重量分析法　gravimetric analysis

转变点　inversion point

转变曲线　inversion curve

转变温度　inversion temperature

状态　state

状态参数　property of state

准静态过程　quasistatic process

自由膨胀　free expansion

自由膨胀过程　free-expansion process

自由膨胀系统　free-expansion system

总温　total pressure

总压　total pressure

总压降　total pressure drop

阻力　resisting force

最大误差　maximum error

最大转变温度　maximal inversion temperature

最大转变压力　maximal inversion pressure

最小转变温度　minimal inversion temperature

最终状态　final state

做功系统　work-producing system

空气动力学

专业术语

B b

白金汉 Pi 定理　Buckingham Pi theorem
北半球　northern hemisphere
背压　back pressure
边界层方程　boundary-layer equations
边界层厚度　boundary-layer thickness
边界层假设　boundary-layer assumption
边界条件　boundary condition
变换　transformation
变形　deformation
标量　scalar
标量场　scalar field
标量积　scalar product
表面力　surface force
表面摩擦　skin friction
表面摩擦阻力　skin friction drag
并行　parallel
并行处理器　parallel processors
伯努利　Bernoulli
伯努利方程　Bernoulli equation
泊松方程　Poisson equation
不可压缩　incompressible
不可压缩流动　incompressible flow
布局　configuration
步长　step size
不稳定　instability

C c

C网格 C-type grids
操纵 maneuver
层流边界层 laminar boundary layer
层流边界层分离 laminar separation
层流-湍流过渡区 laminar-turbulent transition
柴油机 diesel engine
常规布局 conventional configuration
超临界翼 supercritical airfoil
超声速风洞 supersonic wind tunnel
超声速流 supersonic flow
超音速喷管设计 supersonic nozzle design
初始条件 initial conditions
处理器 processor
传导性/传导率 conductivity
垂尾 vertical tail

D d

大黄蜂 bumblebee
大气 atmosphere
单位向/矢量 unit vector
当地导数 local derivative
当地时间步 local time stepping
导热方程 heat conduction equation
导热系数 thermal conductivity
等马赫数爬升 climb at constant Mach number
等熵的 isentropic
等熵过程 isentropic process
等熵状态方程 isentropic equation of state
等效空速 equivalent airspeed

等值线　contour
低密度流　low-density flow
低松弛　underrelaxation
笛卡儿网格　Cartesian grids
笛卡儿坐标　Cartesian coordinates
地面效应　ground effect
地速　ground speed
第二喉部　second throat
第一喉部　first throat
电离层　ionosphere
吊挂　pylon
定常/稳态　steady
定常流体力学　steady fluid dynamics
动量方程　momentum equation
对称翼型　symmetric airfoil
对流层　troposphere
对流层顶层　tropopause
对流项　convective derivative
钝头体　blunt body

E e

二阶中心差分　second-order central difference
二项式展开　binomial expansion

F f

发动机舱　engine nacelle
反射激波　reflected shock
方向导数　directional derivative
方向矢量　position vector
放大因子　amplification factor
飞行包线　flight envelope

非定常/非稳态　unsteady

非定常流体力学　unsteady fluid dynamics

非结构网格　unstructured grids

非守恒型　nonconservation form

非守恒型动量方程　nonconservation form of momentum equation

非守恒型连续性方程　nonconservation form of continuity equation

非守恒型能量方程　nonconservation form of energy equation

分离　flow separation

分离流　separated flow

分离气泡　separation bubble

分子间的　intermolecular

分子间作用力　intermolecular force

风洞　wind tunnel

缝翼　slat

浮力　buoyancy force

辐射加热　radiative heating

附着激波　attached shock waves

附着涡　bound vortex

副翼　aileron

副翼调整片　aileron trim tab

G　g

高超声速流　hypersonic flow

高度计　altimeter

根梢比　taper ratio

固定翼　fixed wing

惯性矩　moment of inertia

滚转　roll

国际标准大气　international standard atmosphere

国际民航组织　International Civil Aviation Organization，ICAO

国家航空宇航局　National Aeronautics and Space Administration，NASA

国家航空咨询委员会　National Advisory Committee on Aeronautics，NACA

过渡流 transition flow

H　h

航程　range

航迹角　flight path angle

耗散　dissipation

喉部　throat

后掠　swecp

后掠角　sweep angle

后缘　trailing edge

后缘失速　trailing-edge stall

后缘襟翼　trailing edge flaps

弧线　camber line

滑翔机　glider

化学反应流　chemically reacting flow

环量　circulation

恢复因子　recover factor

回流　reversed flow

汇流　sink flow

混合长度　mixing length

活塞式发送机　piston engine

火箭发动机　rocket engine

J　j

激波　shock wave

激波捕捉法　shock-capturing method

激波管问题　shock tube problem

激波交点　shock intersection

激波-膨胀波理论　shock-expansion theory

激波脱体距离　shock detachment distance
激波装配法　shock-fitting method
激波阻力　wave drag
积分形式　integral form
机身尾部　empennage
机翼反扭转　washout
机翼根梢比　wing taper ratio
机翼后掠角　wing sweep angle
机翼理论　wing theory
机翼面积　wing area
机翼失速　wing stall
机翼展弦比　wing aspect ratio
机翼正扭转　washin
机翼阻力　wing drag
极曲线　drag polar/lift-drag polar
几何扭转　geometric twist
几何相似　geometric similarity
几何迎角　geometric angle of attack
计算机程序　computer programming
计算机图形技术　computer graphic techniques
计算流体力学　computation fluid dynamics
计算平面　computational plane
迹线　pathline
加力燃烧室　afterburner
加热　heating
加速飞行性能　accelerated flight performance
加速起飞距离　accelerate-go distance
加速停止距离　accelerate-stop distance
剪切力　shear stress
简单襟翼　plain flap
桨盘(螺旋桨)　propeller disk

桨叶　blade
桨叶角　propeller blade angle
交叉方向隐格式/ADI 格式　alternating-direction-implicit technique
交错网格　staggered grids
校准空速　calibrated airspeed
结构网格　structured grids
截断误差　truncation error
解析区域　analytical domain
解向量　solution vector
襟翼　flap
进场　approach
进场高度　approach altitude
进场速度　approach speed
浸湿面积　wetted area
距离　distance
绝对升限　absolute ceiling
绝对温度　absolute temperature
绝热壁面条件　adiabatic wall condition
绝热壁面温度　adiabatic wall temperature
均匀流　uniform flow

K　k

卡门涡街　Karman vortex street
开缝襟翼　slotted flap
开裂式襟翼　split flap
可逆　reversible
可压缩　compressible
可压缩流动　compressible flow
可压缩效应　compressibility effect
可压缩性　compressibility
可用推力　available thrust

空间步长　spatial step size
空间推进　space marching
空气动力　aerodynamic force
空气动力学　aerodynamics
空气动力学的　aerodynamic
空气动力作用中心　aerodynamic center
空速　airspeed
空速管/皮托管　pitot
空速指示器　airspeed indicator
控制点　control point
控制方程　governing equations
控制体　control volume
库塔-儒科夫斯基定律　Kutta-Joukowski theorem
库塔条件　Kutta condition
扩压段/扩压器　diffuser

L l

拉力　pull
拉普拉斯方程　Laplace's equation
拉起　pull-up
来流/自由流　freestream
来流密度/自由流密度　freestream density
来流速度/自由流速　freestream velocity
雷诺平均 N－S 方程　Reynolds-averaged Navier-Stokes，RANS
雷诺应力　Reynolds stress
离散　discretization
离散方程　discretized equations
离散误差　discretization error
离心式压气机　centrifugal flow compressor
理论升限　calculated ceiling
力矩　moment of force

力矩系数　moment coefficient
量纲分析　dimensional analysis
临界马赫数　critical Mach number
零升迎角　zero-lift angle of attack
零升阻力　zero lift drag
零升阻力系数　zero lift drag coefficient
流场　flowfield
流动　flow
流动相似　flow similarity
流动相似性　similarity of flow
流管　streamtube
流函数　stream function
流量/通量　flux
流体静力学方程　hydrostatic equation
流体力学　fluid dynamics
流体微元　fluid element
流线　streamline
流线型外形　steamlined body
龙格-库塔格式　Runge-Kutta scheme
螺旋桨效率　propeller efficiency
螺旋桨性能　propeller performance

M m

马赫反射　Mach reflection
马赫角　Mach angle
马赫数　Mach number
马赫线　Mach line
马蹄涡　horseshoe vortex
面元法　panel technique
摩擦系数　friction coefficient

N n

纳维-斯托克斯方程 Navier-Stokes equation
南半球 Southern hemisphere
内能 internal energy
能量方程 energy equation
逆变换 inverse transformation
逆压梯度 adverse pressure gradient
黏性/黏度 viscosity
黏性的 viscous
黏性耗散 viscous dissipation
牛顿第二定律 Newtonian second law
牛顿理论 Newtonian theory
牛顿流体 Newtonian fluids
扭转 twist

O o

O 网格 O-type grids
欧拉方程 Euler's equation

P p

爬升 climb
爬升率 rate of climb
爬升性能 climb performance
爬升油耗 fuel consumed in climb
抛物型方程 parabolic equations
跑道 runway
配平 trim
喷管/喷嘴 nozzle
喷气发动机 jet engine
膨胀波 expansion wave

偏微分方程 partial differential equation

平板流动 flat plate flow

平飞 level flight

平飞升限 level flight ceiling

平尾 horizontal tail

扑翼/扑翼机 flapping wings

Q q

气动加热 aerodynamic heating

气动扭转 aerodynamic twist

气动弹性变形 aeroelasticity

气动弹性的 aeroelastic

气压高度表 barometric altimeter

启动涡 starting vortex

起飞 take-off

起飞过程 take-off process

起飞距离 take-off distance

起飞距离空中段 take-off air distance

起飞性能 take-off performance

前缘 leading edge

潜艇 submarine

强守恒型 strong conservation form

球体坐标系 spherical coordinate system

曲面积分 surface integral

曲线积分 line integral

R r

燃烧 combustion

燃烧室 combustion chamber

扰动 perturbation

扰流板 spoiler

人工黏性 artificial viscosity
弱守恒型 weak conservation form

S s

萨瑟兰定律 Sutherland's law
三角翼 delta wing
散度 divergence
散度定理 divergence theorem
散逸层 exosphere
刹车 brake
刹车力矩 braking torque
设计俯冲速度 design diving speed
设计升力系数 design lift coefficient
设计巡航速度 design ruising speed
升力 lift
升力面 lifting surface
升力曲线斜率 lift curve slope
升力系数 lift coefficient
升限 ceiling
升致阻力 drag due to lift
升阻比 lift-to-drag ratio
声波 sound wave
声速 velocity of sound
声速流 sonic flow
声速线 sonic line
失速 stall
失速控制装置 stall control devices
失速速度 stalling speed
失速速度 stalling velocity
失速迎角 stall angle
时间步长 time step size

时间推进　time marching
时间相关技术　time-dependent technique
实用升限　practical ceiling/service ceiling
实质导数　substantial derivative
适航　airworthiness
守恒　conservation
守恒型　conservation form
守恒型动量方程　conservation form of momentum equation
守恒型连续性方程　conservation form of continuity equation
守恒型能量方程　conservation form of energy equation
输运现象　transport phenomena
数值耗散　numerical dissipation
数值区域　numerical domain
双曲型方程　hyperbolic equations
双曲型特征　hyperbolic nature
四分之一弦长位置　quarter chord
四阶中心差分　fourth-order central difference
松弛技术　relaxation technique
速度剖面　velocity profile
速度势　velocity potential
速度势方程　velocity potential equation

T　t

TVD 格式　TVD scheme
抬前轮速度　rotation speed
泰勒级数　Taylor's series
弹性　flexibility
特征线　characteristic line
特征向量　eigenvector
特征值　eigenvalue
梯度　gradient

梯度定理　gradient theorem

梯度线　gradient line

梯形　trapezoidal

体积积分　volume integral

调和函数　harmonic functions

调整片　trim tab

贴体网格　boundary-fitted grid

通量矢量分裂　flux-vector splitting

通量限制器　flux limiter

同温层　stratosphere

湍流边界层　turbulent boundary layer

湍流表面摩擦力　turbulent skin friction

湍流动能　turbulent kinetic energy

湍流雷诺数　turbulent Reynolds number

推进解　marching solutions

推进系统　propulsion systems

推力/推进　propulsion

脱体激波　detached shock waves

椭圆形方程　elliptic equation

椭圆形偏微分方程　elliptic partial differential equation

椭圆形特征　elliptic nature

W　w

弯度　camber

完全发展流动　fully developed flow

网格　grid/mesh

网格点　grid point

网格生成　grid generation

微分形式　differential form

尾流　wake

尾涡　trailing vortex

温度剖面　temperature profile
纹影法　schlieren
稳定性判据　stability criterion
涡　vortex
涡导热系数　eddy thermal conductivity
涡环　vortex ring
涡桨发动机　turbo-prop engine
涡流发生器　vortex generator
涡轮　turbine
涡轮发动机　turbine engine
涡喷发动机　turbojet engine
涡破裂　vortex breakdown
涡扇发动机　turbo-fan engine
涡丝　vortex filament
涡黏性　eddy viscosity
无动力下滑　un-powered glide
无滑移条件　no slip condition
无条件稳定　unconditional stability
无旋流　irrotational flow
无黏流动　inviscid flow
无黏性的　inviscid
物理平面　physical plane
误差　error

X　x

下滑/滑翔　glide
下降　descent
下降率　rate of descent
下降性能　descent performance
下洗　downwash
弦　chord

弦线　chord line
显式　explicit form
线性化的　linearized
相对风速　relative wind speed
相似参数　similarity parameters
相似准则　similarity criterion
向后差分　rearward finite difference
向量　vector
向量场　vector field
向量积　vector product
向前差分　forward finite difference
小翼　winglet
斜激波　oblique shock wave
型阻　form drag
需用功率　power required
需用拉力　pull required
悬停　hovering
旋度　curl
旋翼　rotary wing
旋翼机/直升机　rotorcraft
漩涡状态　vorticity
巡航　cruise
巡航速度　cruising velocity

Y y

压差阻力　pressure drag
压力波　pressure wave
压力分布　pressure distribution
压力高度　pressure altitude
压力梯度　pressure gradient
压力系数　pressure coefficient

压力修正公式　pressure correction formula
压力修正技术　pressure correction technique
压力作用中心　center of pressure
压缩率　compressibility
鸭式布局　canard configuration
亚声速　subsonic
亚声速流动　subsonic flow
一阶　first-order
一阶向后差分　first-order rearward difference
一阶向前差分　first order forward difference
一维流动　one-dimensional flow
以相同等效空速爬升　climb at constant equivalent airspeed
翼尖　wingtip
翼尖涡　tip vortex
翼型　airfoil
翼型厚度　airfoil thickness
翼型理论　airfoil theory
翼型最大厚度位置　location of maximum thickness
翼载　wing loading
翼展　wingspan
隐式　implicit methods
迎风差分　upwind difference
迎风格式　upwind scheme
迎角/攻角　angle of attack
油耗　fuel consumed
有限差分　finite difference
有限差分法　finite-difference methods
有限控制体　finite control volume
有限翼　finite wing
有效迎角　effective angle of attack
有效载荷　payload

有旋流　rotational flow
诱导速度　induced velocity
诱导迎角　induced angle of attack
诱导阻力　induced drag
诱导阻力系数　induced drag coefficient
圆柱扰流　circular cylinder flow
圆柱坐标系　cylindrical coordinate system
源流　source flow
源项　source term
远场　far field
运动相似　dynamic similarity
运动黏度　kinematic viscosity

Z z

载荷　loading
载油量　fuel reserves
增升装置　high lift device
展弦比/长细比　aspect ratio
真空速　true airspeed
正激波　normal shock wave
直升机　helicopter
指示空速　indicated air speed
质量力　body force
质量流量　mass flow rate
中弧线　camber line
中心差分　central finite difference
重力加速度　acceleration of gravity
重心　center of gravity
轴流压缩机　axial flow compressor
驻点　stagnation
驻点温度　stagnation temperature

驻点压力　stagnation pressure
转弯　turn
转弯半径　turn radius
状态方程　equation of state
着陆　landing
着陆距离　landing distance
着陆距离空中段　landing air distance
着陆性能　landing performance
自适应网格　adaptive grid
自由分子流　free molecular flow
自由涡　free vortex
总密度　total density
总能　total energy
阻力　drag
阻力系数　drag coefficient
最大升力　maximum lift
最大速度　maximum speed
最大巡航高度　cruise altitude ceiling
最小安全速度　minimum safety speed
最小操控速度　minimum control speed
最小速度　minimum speed
坐标系　coordinate system

飞机结构

专业术语

A a

安定面　stabilizer

B b

钣金件　sheet metal
半硬壳式机身　semi-monocoque fuselage
半硬壳式结构　semi-monocoque structure
棒材　bar
包铝薄板　alclad sheet
包铝层　alclad
薄板　sheet
薄垫片　shim
背鳍　dorsal fin
壁板　panel
标准工艺装备　master tools
表面处理　surface treatment
表面粗糙度　surface roughness
表面光饰　surface finish
箔材　foil
补偿片　compensating plate
不锈钢　stainless steel

C c

操纵面　control surface
槽型件　channel
侧风挡　side windscreen, side windshield

长桁　stringer
超高强度钢　super-high-strength steel
超临界机翼　supercritical wing
超声波探伤　ultrasonic inspection
沉淀硬化不锈钢　precipitation-hardening stainless steel
衬套　bushing
成形　forming
抽芯铆钉　self-plugging rivet
垂尾　vertical tail/wing
垂直安定面　vertical stabilizer
垂直尾翼　vertical tail/fin
磁力探伤　magnetic inspection
淬火　quench
淬火处理　quenching

D d

搭接　overlap splice
带套环的螺纹抽钉　thread smoke nail with lantern ring
倒角　chamfer
登机门　entry door
地板　floor panel
地板梁　floor beam
电子设备舱　avionics bay
垫片　filler，spacer
垫圈　washer
吊挂　pylon
吊挂后梁　pylon rear spar
吊挂后缘整流罩　pylon trailing-edge fairing
吊挂前梁　pylon front spar
吊挂前缘整流罩　pylon leading-edge fairing
钉套　huck collar

动力装置　power plant

镀铬　chromium plate

短舱　nacelle

断裂韧性　fracture toughness

锻件　forging

对接　butt splice

镦头　upset head

E　e

耳片　lug

二维图样　2D drawing

F　f

发动机安装架　engine pylon

发动机舱　engine compartment

发动机短舱　engine nacelle

法兰边　flange

方向舵　rudder

飞机结构　aircraft structure

非金属　nonmetal

非金属的　nonmetallic

风挡　windshield

封闭肋　closing rib

蜂窝　honeycomb

蜂窝夹心　honeycomb sandwich

蜂窝夹心结构　honeycomb sandwich structure

缝翼　slat

服务门　service door

辅助梁　auxiliary spar

复合材料　composite material

副翼　aileron

腹板　web
腹鳍　ventral fin

G g

盖板　cover
干涉配合　interference fit
高锁螺栓　Hi-Lock bolt
高温合金　superalloys
隔框　bulkhead，frame
隔声结构设计　design of structure for sound isolation
铬酸阳极氧化　chromic acid anodize
工具孔　tooling hole
工装协调　tool coordination
工字梁　I beam
公差　tolerance
管材　tubing
滚制波纹　beading
过道　aisle

H h

焊接　welding
合金　alloy
合金钢　alloy steel
盒形翼梁 box spar
桁条　stringer
桁条式结构　stringer structure
横向的　transverse/lateral
后机身　rear fuselage
后缘　trailing-edge，T. E.
后缘襟翼　trailing-edge flap
厚板　plate

化铣蒙皮　chemical-milled skin
化学工艺和表面处理　chemical processes & finishes
化学铣切　chemical milling
环　ring
环槽钉　ring groove nails
回火　temper
回火处理　tempering
锪平　spotfacing

J j

机加操作　machining operation
机加件　machined part
机轮　wheel
机身　fuselage
机身对称中心线　fuselage symmetrical center line
机身基准平面　fuselage reference plane
机头　nose
机翼　wing
基准线　datum line，reference line
挤压强度　bearing strength
挤压型材　extruded section
加劲件　stiffener
加强板　doubler
加强槽　reinforced groove
加强框　reinforced bulkhead/reinforced frame
加强肋　intercostale
驾驶舱　cockpit/flight deck
驾驶舱安全门　cockpit safety door
驾驶舱地板　cockpit floor
间隙配合　clearance fit
减摆器　shimmy damper

减轻孔　lightening hole
减震器　shock absorber
剪切　shear
剪切角材　shear angle
剪切角片　shear clip
剪切应力　shear stress
简单襟翼　plain flap
胶接　adhesive bonding
角材　angle
角度公差　angular tolerance
铰孔　reaming
铰链　hinge
接头　fitting
结构钢　structural steel
金属　metal
金属的　metallic
襟翼　flap
紧配合　tight fit
进气道　air intake
精锻件　precision forging
静不定结构/非静定结构　non-statically determinant structure
静定结构　statically determinant structure

K k

开口补偿设计　cutout structural arrangement design
开口销　cotter pin
可收放式起落架　retractable landing gear
客舱　passenger cabin
客舱地板　cabin floor
框　frame
扩孔　counterboring

L l

拉伸　tension

拉伸角材　tension angle

拉伸应力　tensile stress

雷达罩　radome

梁　spar/beam/longeron

梁式机翼　spar wing

零膨胀设计　design of zero expansion coefficient

硫酸阳极氧化　sulfuric acid anodize

龙骨梁　keel beam

轮毂　hub，boss of wheel

轮胎　tyre

螺钉　screw

螺接　bolting

螺栓　bolt

螺栓头　bolt head

螺纹　thread

螺纹铆钉　rivnut/blind nut

螺桩　stud

铝合金　aluminum alloy

M m

埋头窝　countersink

盲紧固件　blind fastener

盲孔　blind hole

毛坯　blank

铆钉　rivet

铆钉头　rivet head

铆接　riveting

帽型材　hat (section)
镁合金　magnesium alloy
蒙皮　skin
模锻件　molded forging
模线　lofting
磨损　abrasion

N　n

耐腐蚀钢　corrosion resistant steel
耐腐蚀设计　corrosion-proof design
内圆角半径　fillet radii
扭矩　torque
扭转　torsion

P　p

盘型件 pan
配合　fit
疲劳　fatigue
偏差　deviation
拼接　splice，splicing
平尾　horizontal tail (wing)
平直机翼　straight wing

Q　q

起落架　landing gear，undercarriage
起落架舱门　landing gear door
起落架轮舱　landing gear wheel well
气密舱　pressurized cabin
气密框　pressure bulkhead
前机身　forward fuselage

前起落架　nose landing gear，NLG

前三点起落架　tricycle landing gear

前缘　leading-edge，L. E.

前缘缝翼　leading-edge slat

前缘襟翼　leading-edge flap

球面框　hemispherical bulkhead

全机身　whole fuselage

R　r

扰流板　spoiler

热处理　heat treatment

韧性　toughness

S　s

三维数模　3D mathematical model

刹车装置　brake device

设备舱门　equipment door

升降舵　elevator

声疲劳　acoustic fatigue

实心铆钉　solid rivet

视界　visibility

竖线　buttock line

数学模型　mathematical model

数字化样机　digital mockup，DMU

数字化预装配　digital pre-assembly，DPA

双腔起落架　landing gear with a two stage shock absorber

水平安定面　horizontal stabilizer

水平尾翼　horizontal tail

水线　water line

松配合　loose fit

T t

T 形尾翼　T-tail

T 型材　T-section

钛合金　titanium alloy

碳钢　carbon steel

镗孔　boring

填角　fillet

条带　strap

通风窗　aeration window

涂层　coating

退火　anneal

退火处理　annealing

托板螺母　nut plate

W w

外螺纹　external thread

弯矩　bending moment

弯曲　bending

弯曲半径　bending radius

弯曲角　bending angle

弯曲线　bending line

维护口盖　access panel/access door

尾翼　tail unit

涡扇发动机　turbofan engine

无内胎轮胎　tubeless tyre

无损探伤　non-destructive inspection

无头铆钉　slug

X x

吸声结构设计　design of structure for sound absorption

下陷　joggle/sink

先进复合材料　advanced composite

舷窗　cabin window

线材　wire

线性公差　linear tolerance

消声结构设计　design of structure for noise elimination

销钉孔　pin hole

斜角　bevel

斜梁　diagonal spar

泄压门　vent door

行李舱　luggage compartment/baggage compartment

型材　section/shape/profile

型架　jig

Y y

压缩　compression

压缩应力　compression stress

阳极氧化　anodize

阳极氧化处理　anodize treatment/anodizing

样板　template

摇臂式起落架　levered suspension landing gear/trailing link landing gear

翼尖　wing tip

翼梁　wing spar

翼梢小翼　winglet

翼箱　wing box

应急舱门　emergency door

应急出口　emergency exit

硬度　hardness

硬壳式机身　monocoque fuselage

硬质阳极氧化　hard anodize

油箱　fuel tank

缘条(凸缘)　flange/cap

Z z

增升装置　high lift device

增压座舱　pressurized cabin/sealed cabin

站位　station

站线　station line

整流罩　fairing

整体油箱　integral fuel tank

正面进气道　dorsal air intake/front air intake

支架　bracket

支柱　strut

支柱式起落架　telescopic landing gear

中后机身　mid-aft fuselage

中前机身　mid-fwd fuselage

轴承　bearing

主风挡　windscreen/windshield

主起落架　main landing gear

柱　column

铸件　casting

转弯操纵机构　wheel steering system

锥形螺栓　taper-lok/tapered bolt

着色检查　penetrant inspection

自由锻件　hand forging

纵向的　longitudinal

钻孔　drilling

最小壁厚　minimum wall thickness

座舱盖　cockpit cover

座舱声学设计　acoustic design of cabin

座椅滑轨　seat track

环境控制系统

导 语

环境控制系统通常是指气源系统和空调系统。

气源系统的主要功能是控制来自发动机、APU或者地面高压气源的高温高压空气的温度、压力和流量，为空调系统和其他用气系统（如防冰和发动机起动）提供气源。气源系统根据构型进行气源选择的控制（发动机引气、APU引气或者是地面高压气源车），并通过对交叉引气活门的控制，实现左右两侧引气隔离。

空调系统包括制冷、空气分配、座舱增压控制和电子设备冷却子系统。空调系统主要为驾驶舱和旅客舱提供温度、压力和湿度适宜的调节空气。制冷子系统通常设置两套空调制冷组件来降低来自气源系统的气体温度，同时除掉空气中多余的游离水。冲压空气用来对空调制冷组件的热空气进行冷却，当制冷组件故障时也可为空气分配系统提供应急通风。制冷子系统通常通过组件出口压力传感器和组件出口温度传感器分别对出口空气的温度和压力进行监控。空气分配子系统通过合理的管网和流道设计对供入座舱的空气进行分配。大部分的民用飞机会设置再循环子系统来减小进入旅客舱区域的供气温差。通常，电子设备风扇抽取舱内排气对机载电子设备和驾驶舱显示屏进行通风冷却，有时也会采用液体冷却系统来冷却机载电子设备。在地面运行及全部飞行状态中，座舱压力调节系统提供一个安全舒适的座舱压力环境，并对座舱增压参数进行指示、告警等。通常，座舱压力调节系统通过对座舱排气活门进行调节（以飞行剖面和飞机构型的函数形式）实现座舱压力的自动控制，手动控制作为备份，利用安全活门实现座舱的正压和负压保护。地面状态时，利用地面活门限制座舱内外压差。

Environment control system (ECS) mainly refers pneumatic system and air conditioning system .

The main function of pneumatic system (PNU) is to control the temperature, flow and pressure of high pressure air, which is extracted from engine and also from alternate sources such as the APU and high pressure ground cart. The PNU provides air sources for air conditioning system (ACS) and other users system such as ice protection system and engine start. The PNU manages air source selection (engine or APU bleed air or high pressure ground cart) according to the configuration control. Otherwise, The PNU manages left hand side or right hand side isolation by controlling and monitoring a cross bleed valve.

The ACS includes the subsystems that following: cooling, distribution, cabin pressurization control and avionics cooling. The function of the air conditioning system is to supply the passengers and flight crew with conditioned air at a comfortable temperature, pressure and humidity level. Two air conditioning units are always designed in cooling subsystem to decrease the temperature of the air from the pneumatic system, and remove water from the air at the same time. Ram air is used to cool the air conditioning units and may also be used to supply emergency air to the distribution subsystem, if the air conditioning units are not operating. The cooling subsystem generally monitors the pressure and temperature of the discharged air by pack discharge pressure sensor and pack discharge temperature sensor separately. The air distribution subsystem channels temperature-controlled air to the flight deck and passenger cabin through a network of distribution ducts. A recirculation subsystem is usually designed in passenger compartment to decrease the temperature difference of supply air in most civil airplane. Generally, exhaust air of cabin is extracted by electric fans to cool avionics and cockpit display. Liquid cooling system is also utilized for avionics cooling sometimes. The cabin pressure control system (CPCS) provides maximum safety and comfort environment, as well as cabin pressure indication and warning, during all segments of flight and ground operations. Normally, cabin pressurization is controlled and scheduled automatically (as a function of the aircraft profile and configuration) by a cabin pressure controller with provisions for manual backup control through regulating cabin outflow valve. Safety valves prevent excessive positive or negative pressures in the cabin. A ground valve is used to limit differential pressure on the ground.

缩略语

𝒜 𝒶

A/C　Aircraft　飞机
AC　Alternating Current　交流电
ACM　Air Cycle Machine　空气循环机
ACS　Air Conditioning System　空气调节系统
ACU　Air Conditioning Unit　空气调节组件
AEV　Avionics Exhaust Valve　电子设备排气活门
AMS　Air Management System　空气管理系统
APU　Auxiliary Power Unit　辅助动力装置
ARINC　Aeronautical Radio Inc　航空无线电
ATA　Air Transport Association of America　美国航空运输协会
AVFAN　Avionics Cooling Supply Fan　电子设备冷却风扇

ℬ 𝒷

BAS　Bleed Air System　引气系统
BBHEAT　Baggage Bay Heater　货舱加热器
BBSOV　Baggage Bay Shut-Off Valve　货舱关断活门
BCKV　Bulkhead Check Valve　隔框单向活门
BFR　Bleed Filter　引气过滤器
BIT　Built In Test　机内自检测
BTS　Bleed Temperature Sensor　引气温度传感器

𝒞 𝒸

CAD　Computer Aided Design　计算机辅助设计
CAS　Crew Alerting System　乘务员告警系统
CBV　Cross Bleed Valve　交叉引气活门

CDT　Compressor Discharge Temperature　压气机排气温度

CKV　Check Valve　单向活门

CMC　Centralized Maintenance Computer　中央维护计算机

CMD　Command　命令

CMM　Component Maintenance Manual　设备维修手册

CMS　Centralized Maintenance System　中央维护系统

COND　Condenser　冷凝器

CPCP　Cabin Pressure Control Panel　座舱压力控制面板

CPCS　Cabin Pressure Control System　座舱压力控制系统

CPU　Central Processing Unit　中央处理单元

D　d

DAU　Data Acquisition Unit　数据接收单元

DB　Dry Bulb　干球

DC　Direct Current　直流

DCU　Data Control Unit　数据控制单元

DHX　Dual Heat Exchanger　两级热交换器

DTS　Duct Temperature Sensor　管路温度传感器

E　e

ECS　Environment Control System　环境控制系统

EE　Electronic Equipment　电子设备

EICAS　Engine Indication and Crew-Alerting System　发动机指示和乘务员告警系统

EMI　Electromagnetic Interference　电磁干扰

F　f

FAV　Fan Air Valve　风扇空气活门

FC　Full Closed　全关

FCV　Flow Control Valve　流量控制活门

FO　Full Open　全开

FMEA　Failure Modes&Effects Analysis　故障(失效)模式及影响分析

FMECA　Failure Modes Effects and Criticality Analysis　故障(失效)模式影响及危害性分析

FSV　Flow Sensor Venturi　文氏管流量传感器

FTA　Fault Tree Analysis　故障树分析

FWD　Forward Part of Cabin　座舱前部

FWS　Failure and Warning System　失效告警系统

G g

GHEAT　Galley Heater　厨房加热器

GND　Ground　地面

GSE　Ground Support Equipment　地面保障设备

GV　Ground Valve　地面活门

H h

H/W　Hardware　硬件

HACKV　Hot Air Check-Valve　热空气单向活门

HP　High Pressure　高压

HPGC　High Pressure Ground Connection　高压地面接头

HPV　High Pressure Valve　高压活门

HX　Heat Exchanger　热交换器

I i

IASC　Integrated Air System Controller　空调系统综合控制器

IBIT　Initiated Built in Test　启动机内自检测

IOC　Input/Output Controller　输入输出控制器

IP　Intermediate Pressure　中压

IPCV　Intermediate Pressure Check Valve　中压单向活门

IPL　Illustrated Parts List　图解零部件目录

ISA　International Standard Atmosphere　国际标准大气

L l

L/H　Left Hand　左侧

LCS　Liquid Cooling System　液体冷却系统

LCV　Load Control Valve　载荷控制活门

LP　Low Pressure　低压

LPGC　Low Pressure Ground Connection　低压地面接头

LRU　Line Replaceable Unit　航线可更换单元

M m

MHX　Main Heat Exchanger　主热交换器

MIXM　Mix Manifold　混合腔

MMTS　Mixing Manifold Temperature Sensor　混合腔温度传感器

N n

NFC　Not Full Close　非全关

NFO　Not Full Open　非全开

NTU　Number of Heat Transfer Unit　传热单元数

O o

OAT　Outside Air Temperature　外界空气温度

OC　Open Circuit　开环

OFV　Outflow Valve　排气活门

OZC　Ozone Conversion　臭氧转换器

P p

PB　Push Button　按钮

PBIT　Power Up Built Test　上电自检测

PCE　Precooler Exchanger　预冷器

PDT　Pack Discharge Temperature　组件出口温度

PDTS　Pack Discharge Temperature Sensor　组件出口温度传感器

PDP　Pack Discharge Pressure　组件出口压力
PDPS　Pack Discharge Pressure Sensor　组件出口压力传感器
PHX　Primary Heat Exchanger　初级热交换器
PIFS　Pack Inlet Flow Sensor　组件入口流量传感器
PIP　Pack Inlet Pressure　组件入口压力
PIPS　Pack Inlet Pressure Sensor　组件入口压力传感器
PIT　Pack Inlet Temperature　组件入口温度
PITS　Pack Inlet Temperature Sensor　组件入口温度传感器
PL　Plenum　充压
PNU　Pneumatic System　气源系统
PRSOV　Pressure Regulating and Shut-Off Valve　压力调节关断活门
PS　Pressure Sensor　压力传感器
PTS　Pack Temperature Sensor　组件温度传感器

R r

R/H　Right Hand　右侧
RAV　Ram Air Ventilation Valve　冲压空气通风活门
RECF　Recirculation Filter　再循环过滤器
REH　Reheater　回热器
RFANR　Recirculation Fan　再循环风扇

S s

S/W　Software　软件
SL　Sea Level　海平面
SOV　Shut-Off Valve　关断活门
SV　Safety Valve　安全活门

T t

T Duct　Telescopic Duct　伸缩管
TAPRV　Trim Air Pressure Regulate Valve　配平空气压力调节活门
TAV　Trim Air Valve　配平空气活门

TBC To Be Confirmed 待确认

TBD To Be Defined 待定

TCT Temperature Control Thermostat 温度控制调节器

TCV Temperature Control Valve 温度控制活门

V v

VENTS Ventilated Temperature Sensors 通风温度传感器

VPC Cabin Pressure Rate 座舱压力变化率

VPLIM Cabin Pressure Rate Limitation 座舱压力变化率限制值

VPLIMC Cabin Pressure Rate Limitation in Climb 爬升模式座舱压力变化率限制值

VPLIMD Cabin Pressure Rate Limitation in Descent 下降模式座舱压力变化率限制值

VZA Aircraft Vertical Rate 飞机垂直速度

W w

WBGT Web Bulb Globe Temperature 三球温度/湿球黑球温度

WE Water Extractor 水分离器

WS Water Sprayer 喷水器

WOW Weight on Wheel 轮载

Z z

ZA Aircraft Altitude 飞机飞行高度

专业术语

A a

安全/释放阀　relief valve

B b

百分比湿度　percentage humidity

板翅通道　flow passage of plate fin

板束　plate bundle

半开式叶轮　half opening impeller

饱和　saturation

爆破压力　burst pressure

泵　pump

比定容热容　specific heat capacity at constant volume

比定压热容　specific heat capacity at constant pressure

比湿度/湿度比　specific humidity/humidity ratio

闭环　closed loop

闭式空气循环制冷系统　closed air-cycle refrigeration system

闭式循环　closed cycle

闭式叶轮　closed impeller

变容积式飞机环境控制系统　aircraft environmental control system with positive displacement air-cycle machine

变容积压气机　positive displacement compressor

标准大气　standard atmosphere

标准高度　standard altitude

标准海平面大气　standard sea level air

标准流量　standard flow

表面放热系数　surface or film heat transfer coefficient

表面效率　surface effectiveness

表面张力　surface tension

波动　fluctuate

波纹管　bellows

波纹管的有效面积　effective area of bellows

不可恢复总压降　total nonrecoverable pressure drop

C　c

材料　materials

侧板　side plate

超温　overtemperature

超压　overpressure

齿轮泵　gear pumps

翅片　fin

翅片长度　fin length

翅片高度　fin height

翅片间距　fin separation

翅片面积　fin area

翅片效率　fin effect

充填　charge

冲压空气　ram air

冲压空气温升　ram air temperature rise

出口压力损失　outlet pressure loss

出口压力损失系数　outlet pressure loss coefficient

初级热交换器　primary heat exchanger

厨房盥洗室通风系统　galley and lavatories ventilation system

储液箱　reservoir

传热面积　heat transfer area

喘振边界　surge limit

喘振裕度　surge margin

串联制冷系统　cascade refrigeration system

次级热交换器　secondary heat exchanger

𝒟　*d*

大气温度　atmosphere temperature

单向活门　check valve

挡油衬套　slinger

导管　duct

导热率/导热系数　thermal conductivity

等熵压缩功　isentropic compression work

等熵压缩过程　isentropic compression process

等温区　isothermal region

等温压缩过程　isothermal compression process

等效高度　equivalent altitude

等压控制　isobaric control

低压除水简单式空气循环制冷系统　Simple-cycle refrigeration system with separate low-pressure water separator

低压除水三轮式空气循环制冷系统　Three-wheel bootstrap-cycle refrigeration system with separate low-pressure water separator

低压除水升压式空气循环制冷系统　Bootstrap-cycle refrigeration system with separate low-pressure water separator

低压水分离器　low pressure water separator

电加热器　electric heater

电子-电动式座舱压力调节器　electro-electric cabin pressure regulator

电子-气动式座舱压力调节器　electro-pneumatic cabin pressure regulator

定性温度　qualitative temperature

毒性　toxicity

对流层/平流层　troposphere/substratosphere

对流平均温差　logarithmic - mean temperature difference

多变压缩功　polytropic compression work

多变压缩过程　polytropic compression process

F f

法兰　flange

反作用度　degree of reaction

防喘振装置　surge-preventing device

防护涂层　protective coating

防火材料　fire-proof materials

放热系数 heat emission coefficient

飞机性能代偿损失　aircraft performance penalty

飞行前空气调节　preflight air conditioning

非增压座舱　nonpressurized cabin

废热空气加热器　exhaust hot air type heater

沸腾器　boiler

分贝　decibel

分压　partial pressure

风扇　fan

风扇消耗功率　fan expense power

风扇效率　fan efficiency

风扇有效功率　fan useful power

封头　header

辐射式加热器　radiant heater

辅助进气口　auxiliary air intake

腐蚀防护　corrosion protection

负压　negative pressure

负压区　negative pressure area

G g

干空气　dry air

干空气计算温度　dry air rated temperature

干球温度　dry bulb temperature

干燥　dehydration

干燥剂　desiccant
干燥器　dryer
刚性转子　rigid rotor
高度传感器　altitude transducer
高压除水回流简单式空气循环制冷系统　simple-cycle refrigeration system with recirculation-type high-pressure water separator
高压水分离器　high pressure water separator
隔板　partition plate
隔热　thermal insulation
隔声　sound insulation
个别通风喷头　individual gasper air outlet
个别通风系统　individual air distribution system
工作压力　operating pressure
功率系数　power coefficient
共用传热面积　common heating surface area
故障隔离　fault isolation
观察窗　sight class
管板　tube plate
管束　tube bundle
硅胶　silica gel
滚珠轴承涡轮冷却器　ball bearing turbine cooling
过渡端盖　end shroud
过冷　subcooler
过滤器　filter
过热蒸汽　superheated vapor

H h

含湿量　humidity ratio
恒温器　thermostat
滑油充填量　oil charge
环境温度　ambient temperature

换热管　heat exchanger tube
换热量　heat transfer rate
换热器效率　heat exchanger efficiency
回环式冷却系统　shoestring cooling system
混合腔　mix manifold
活性炭　activated carbon

J j

级　stage
级膨胀比　turbine expansion ratio
加/排液接头　fill/drain connection
加热系统　heating system
驾驶舱　cockpit
间接式座舱压头调节器　indirect cabin pressure controller
间隙漏气损失　gap leakage loss
间隙漏气损失系数　gap leakage energy loss coefficient
减振器　damping device
简单/升压式冷却系统　simple/bootstrap cooling system
简单冷却系统　simple cooling system
简单式空气循环制冷系统　simple cycle refrigeration system
降低环境温度的冷却系统　reduced ambient cooling system
校准流量　calibrated flow
节流　throttling
紧凑系数　compactness factor
进口压力损失　inlet pressure loss
进口压力损失系数　inlet pressure loss coefficient
进气系统　induction system
径流式涡轮冷却器　radial-flow turbine cooling
径向速度　radial velocity
静压差　static pressure differential
局部阻力系数　local drag loss coefficient

绝对速度　absolute velocity
绝对压力调节器　absolute pressure regulator
绝对黏度　absolute viscosity
绝热指数　adiabatic exponent

K k

壳体　casing
可调喷嘴涡轮冷却器　variable nozzle turbine cooling
可靠性　reliability
可燃性　flammability
空调系统气源　air conditioning supply
空空热交换器　air-to-air heat exchanger
空气壁传热率　air space conductance
空气出口　air outlet
空气调节　air conditioning
空气分配　air distribution
空气过滤器　air filter
空气回流　air circulation
空气进口　air inlet
空气流量调节器　air flow regulator
空气喷头　air outlet gasper
空气循环制冷系统　air-cycle refrigerating system
空气轴承　air bearing
空液热交换器　air-to-liquid heat exchanger
孔度　porosity
快卸　quick disconnect
扩压器　diffuser

L l

冷板　cold plate
冷吨　ton of refrigeration

冷凝器　condenser

冷却空气　cooling air

冷却效果检测器　cooling effect detector

冷却液　liquid coolant

冷却装置　chiller

离肖尔姆式/罗宋式压气机　Lysholm type compressor

离心泵　centrifugal pump

离心式分离器　centrifugal separator

离心式风扇　centrifugal fan

离心式压气机　centrifugal compressor

理论速度　theoretical velocity

两轮升压式空气循环制冷系统　two-wheel bootstrap cycle refrigeration system

临界点　critical point

临界转速　critical speed

淋雨　rain

流量系数　flow coefficient

露点温度　dew point temperature

轮径比　diameter ratio

轮盘摩擦能量损失系数　impeller friction energy loss coefficient

轮盘摩擦损失功率　impeller friction loss power

轮轴效率　wheel efficiency

罗茨式压气机　roots-type compressor

螺旋密封　screw seal

M　m

毛细管　capillary tube

霉菌　fungus

迷宫式密封　labyrinth packing

密封　seal

摩擦阻力　friction drag

摩擦阻力系数　friction drag coefficient

膜片　diaphragm

膜盒　capsule

膜盒有效面积　effective area of capsule

N　n

耐火材料　fire-resistant materials

耐燃材料　flame-resistant materials

耐闪材料　flash-resistant materials

耐压压力　proof pressure

挠性转子　flexible rotor

内能　internal energy

内燃式加热器　internal combustion type heater

内套管　tube intensifier

逆流换热　countercurrent flow

逆升压式空气循环制冷系统　reversed bootstrap air-cycle refrigeration system

凝露　condensation

努塞尔数　Nusselt number

P　p

排气系统　exhaust system

排泄活门　expansion valve

旁路活门　by-pass valve

配平空气　trim air

喷嘴　nozzle

喷嘴出口气流角　nozzle flow outlet angle

喷嘴喉部宽度　nozzle throat width

喷嘴喉部面积　nozzle throat area

喷嘴环　nozzle ring

喷嘴环能量损失　nozzle ring energy loss

喷嘴环能量损失系数　nozzle ring energy loss coefficient
喷嘴几何面积　geometric nozzle area
喷嘴气流偏转角　nozzle flow deflection angle
喷嘴叶片安装角　nozzle blade stagger angle
喷嘴有效面积　effective nozzle area
喷射泵/引射器　jet pump/ejector
膨胀阀　expansion valve
膨胀箱　expansion box

Q　q

起调高度　initiation altitude
气动式座舱压力调节器　pneumatic cabin pressure regulator
气体轴承涡轮冷却器　gas bearing turbine cooling
气隙传热率　air space conductance
气压控制　barometric control
潜热　latent heat
潜热热载荷　latent heat load
强迫对流　forced convection
切向速度　tangential velocity
球形感温泡　thermostatic bulb
全流量高压除水三轮式空气循环制冷系统　three-wheel bootstrap-cycle refrigeration system with full-flow high-pressure water separator
全流量高压除水升压式空气循环制冷系统　bootstrap-cycle refrigeration system with full-flow high-pressure water separator

R　r

燃油-空气换热器　fuel-air heat exchanger
热沉　heat sink
热电冷却　thermoelectric cooling
热辐射　thermal radiation

热管　heat pipe

热交换器　heat exchanger

热交换器芯体　heat exchanger core

热交换效率　heat exchanger effectiveness

热绝缘系数　thermal insulance coefficient/coefficient of thermal insulation

热力学　thermodynamic

热膨胀　thermal expansion

热源　heat source

热载荷　heat load

热阻　thermal resistance/heat resistance

熔点/冰点　pour point/freezing point

柔性接头　flexible connection

S s

三轮升压式空气循环制冷系统　three-wheel bootstrap cycle refrigeration system

三轮式涡轮冷却器　three wheel turbine cooling

沙尘控制　dust control

闪点　flash point

烧结式过滤器　sintered filter

设备热载荷　equipment heat load

设计温度　design temperature

渗透热载荷　infiltration heat load

升压式冷却系统　bootstrap cooling system

声强　sound intensity

声强级　sound intensity level

湿度调节器　humidity regulator

湿度控制　humidity control

湿球温度　wet bulb temperature

湿热　humid heat

湿蒸汽干度　quality of wet vapor

实际速度　real velocity

手动控制　manual control

水当量　water equivalent

水当量比　water equivalent ratio

水当量直径　hydraulic diameter

水分控制　water control

水分离器　water separator

水分离效率　water separator efficiency

水蒸发器　water evaporator

水蒸气压　water vapor pressure

速度三角形　velocity triangle

速度系数　velocity coefficient

速压　velocity pressure

算术平均温差　arithmetic-mean temperature difference

算术平均温度　arithmetic-mean temperature

T t

太阳辐射　solar radiation

太阳辐射热载荷　solar heat load

套筒式加热器　muff type heater

体积流量　volume flow rate

调制控制　modulating control

通风　ventilating

同温层　stratosphere

推力径向轴承　thrust-journal bearing

推力轴承　thrust-bearing

W w

外壳　shell

往复式压气机　reciprocating compressor

维修性　maintainability
温度补偿器　temperature compensator
温度冲击　temperature shock
温度传感器　temperature sensor
温度调节器　temperature regulator
温度恢复系数　temperature recovery coefficient
温度继电器　temperature switch
温度控制盒　temperature controller
温度控制系统　temperature control system
温度选择器　temperature selector
温湿指数　temperature humidity index
文氏管　venturi
涡流管　vortex tube
涡轮比直径　specific diameter
涡轮比转速　specific speed
涡轮等熵焓降　isentropic enthalpy drop of turbine
涡轮-风扇式涡轮冷却器　turbine-fan turbine cooling
涡轮绝热功率　turbine adiabatic power
涡轮绝热效率　turbine adiabatic efficiency
涡轮冷却器转速调节器　speed controller of turbine cooling unit
涡轮冷却系统　turbine cooling system
涡轮理论温降　turbine theoretical temperature drop
涡轮膨胀比　turbine expansion ratio
涡轮实际温降　turbine actual temperature drop
涡轮速比　velocity ratio
涡轮特性曲线　turbine characteristic curve
涡轮蜗壳　turbine volute casing
涡轮-压气机式涡轮冷却器　turbine-compressor turbine cooling
涡轮叶轮　turbine impeller
涡轮有效功　turbine useful power
涡轮有效功率　turbine useful efficiency

涡轮有效效率　turbine useful efficiency
无因次特性曲线　dimensionless characteristic curve

X　x

吸附　absorption
吸附剂　absorbent
系统构型　system configuration
系统性能系数　coefficient of performance for system
显热　sensible heat
显热热载荷　sensible heat load
限流器　choke
相对湿度　relative humidity
相对速度　relative velocity
泄漏量　leakage rate
芯体　core
性能系数　coefficient of performance，COP
休止层　tropopause
修正后温度　corrected temperature
旋转式换热器　rotating heat exchanger

Y　y

压比　pressure ratio
压差控制　differential pressure control
压差转换　differential pressure conversion
压力比调节器　pressure ratio regulator
压力比控制　pressure ratio control
压力传感器　pressure transducer
压力恢复　pressure recovery
压力脉动　pressure pulsation
压气机　compressor
压气机喘振　compressor surging

压气机多变效率　compressor polytropic efficiency
压气机绝热系数　compressor adiabatic efficiency
压气机特性曲线　compressor characteristic curve
压气机有效功　compressor useful work
压气机总耗功　compressor total expense work
压头系数　coefficient of head
盐雾　salt mist
叶高　blade height
叶轮出口气流绝对速度角　impeller flow outlet absolute velocity angle
叶轮出口气流相对速度角　impeller flow outlet relative velocity angle
叶轮进口气流绝对速度角　impeller flow inlet absolute velocity angle
叶轮进口气流相对速度角　impeller flow inlet relative velocity angle
叶轮能量损失　impeller energy loss
叶轮能量损失系数　impeller energy loss coefficient
叶片　blade
叶片式压气机　vane compressor
叶型　blade profile
叶型中线　blade axis
叶栅　cascade
液体传输系统　liquid transport system
液位指示器　fluid level indicator
翼弦　chord
迎风面积　face area/windward area
引气关断活门　bleed air shut-off valve
引气系统　bleed air system
引射器　ejector
应急通风　emergency ventilation
油箱　fuel tank
油芯　oil-wick
有效传热面积　efficient heat transfer surface
余速能量损失　leaving velocity energy loss

预感器　anticipator
预冷器　precooler heat exchanger
圆周速度　peripheral velocity
运动黏度　kinematic viscosity

Z　z

再生式换热器　regenerative heat exchanger
再生式冷却系统　regenerative cooling system
再循环空气　recirculated air
噪声　noise
增湿器　humidifier
增压比　pressure ratio of compressor
增压座舱　pressurized cabin
蒸发器　evaporator
蒸发循环过冷冷却器　vapor cycle subcooler
蒸发循环制冷系统　vapor cycle refrigeration system
蒸发循环贮液器　vapor cycle liquid receiver
正排量压气机　positive displacement compressor
正压区　positive pressure area
支架　bracket
直接式座舱压力调节器　direct cabin pressure regulator
制冷包　cooling pack
制冷剂　refrigerant/coolant
制冷剂充填量　refrigerant charge
制冷量　refrigerant
制冷能力　cooling capacity
制冷系统　refrigerating system
制冷循环　refrigeration cycle
质量流量　mass flow rate
质量密度　mass density
周速系数　coefficient of peripheral velocity

轴承壳体　bearing casing
轴颈轴承　journal-bearing
轴流式涡轮冷却器　axial-flow turbine cooling
轴流式压气机　axial compressor
轴向速度　axial velocity
主轴　main shaft
转子　rotor
自动控制　automatic control
自密封　self sealing
自然对流　free convection
自由流通面积　free flow area
总传热系数　overall heat transfer coefficient
总加热载荷　total heating load
总冷却载荷　total cooling load
总泄漏量　total leakage
总压　total pressure
总压差　total pressure differential
阻尼器　damper
阻塞系数　coefficient of choke
组合式制冷系统　compound refrigeration system
最大载荷　peak load
座舱　cabin
座舱安全活门　cabin safety valve
座舱防雾和防霜系统　cabin anti-fog and anti-frost system
座舱负压活门　cabin pressure inflow safety valve
座舱高度　cabin altitude
座舱供气量　cabin air supply slow rate
座舱加热器　cabin heater
座舱加热载荷　cabin heating load
座舱绝对压力　cabin absolute pressure
座舱空气分配系统　cabin air distribution system

座舱空气换气次数　times of cabin air exchange
座舱冷却载荷　cabin cooling load
座舱排气活门　cabin outflow valve
座舱湿度调节系统　cabin humidity control system
座舱通风　cabin ventilation
座舱温度调节系统　cabin temperature control system
座舱消声器　cabin silencer
座舱泄漏量　cabin leakage rate
座舱压力变化率　cabin pressure change rate
座舱压力变化速率限制器　cabin pressure change rate limiting device
座舱压力调节器　cabin pressure regulator
座舱压力调节系统　cabin air pressure control system
座舱压力调节系统动态特性　dynamic characteristic for cabin pressure control system
座舱压力调节系统静态特性　static characteristic for cabin pressure control system
座舱压力辅助调节器　cabin pressure auxiliary unit regulator
座舱压力制度　cabin pressure schedule
座舱压力主调节器　master cabin pressure regulator
座舱应急释压　emergency release of cabin pressure
座舱应急卸压活门　emergency release valve of cabin pressure
座舱余压　cabin pressure difference
座舱增压　cabin pressurizing
座舱增压器　cabin compressor
座舱增压中间冷却器　cabin pressurizing intercooler
座舱最小压力　cabin minimum pressure

防冰除雨系统

导　语

防冰系统用以保证飞机在结冰气象条件下的安全飞行和操纵。防冰系统由 6 个子系统组成，分别是机翼防冰系统、发动机进气道防冰系统、探测器防冰系统、风挡防冰系统、风挡除雨系统、结冰探测系统。

机翼热气防冰系统采用发动机引气防止机翼前缘结冰。系统由机翼防冰活门、压力开关、伸缩管、补偿套管、笛形管、机翼和发动机防冰控制面板、测试面板组成。

发动机进气道热气防冰系统利用发动机引气防止发动机进气道唇口结冰。发动机进气道系统由发动机进气道防冰活门、发动机短舱唇口防冰管路、机翼和发动机防冰控制面板组成。

在探测器防冰系统中，动压管与静压管、总温空气传感器与迎角传感器都是通过电加热防止结冰来保证传感器的准确度。电子发动机控制传感器通过电加热防止结冰来保证传感器的准确度。传感器加热是自动控制的。

风挡加温系统防止风挡在结冰条件下发生结冰，或者除掉风挡表面的冰，以确保飞行员视野不受限制。通常采用电加热元件的方式进行加热，确保风挡表面温度足以防冰但不会损害玻璃结构。

风挡除雨系统协助飞行员在雨雪天视野清晰。风挡除雨系统包括风挡雨刷系统与除雨剂系统。

结冰探测系统提醒飞行员或者飞机系统(或两者)有结冰条件存在，或者飞机表面上已经结冰。结冰探测系统可以是主动式的或者咨询式的。

The anti-icing system makes sure the airplane systems operate correctly when ice occurs. The anti-icing system includes six subsystems：wing thermal anti-icing，engine inlet thermal anti-icing，probe anti-icing，flight compartment window anti-icing，and ice detection system.

The wing thermal anti-icing (TAI) system uses engine bleed air to prevent ice on the leading edge slats. The system components include the wing TAI valves, wing pressure switches, telescoping ducts, flexible couplings, spray tubes, wing and engine anti-ice control panel, and test panel.

The engine icinlet TAI system uses engine bleed air to prevent ice on the engine inlets. The system components include the engine inlet TAI valves, engine inlet TAI pressure switches, engine inlet TAI ducts, and wing and engine anti-ice control panel.

In probe anti-icing system. the pitot-static probes, total air temperature probes, and angle of attack sensors are electrically heated to prevent the formation of ice which would affect sensor accuracy. The Electronic Engine Control (EEC) PT2/TT2 probes are electrically heated to prevent the formation of ice which would affect sensor accuracy。Probe heat is controlled automatically.

Windshield heating systems prevent, or to clear accumulations of, ice from the windshields during icing conditions to ensure pilots' field of vision is not restricted during flight into icing conditions. These surfaces are usually protected by electric heat elements. It confirms that the windshield surface temperature is sufficient to prevent ice accumulation without causing structural damage to the windshield.

The rain protection system helps the pilots see better during rain or snow. The rain protection system includes the windshield wiper system and the rain repellent system.

Icing Conditions Detection Systems alert the flight crew or aircraft systems (or both) to the presence of icing conditions or ice buildup on aircraft surfaces. These ice detection systems may be primary or advisory.

缩略语

A a

ADF　Anti-icing/De-icing Fluid　防冰/除冰液

AFIDS　Advisory flight ice detector system　咨询式结冰探测系统

AI　Anti-icing　防冰

AICKV　Anti-icing Check Valve　防冰单向活门

AIDPS　Anti-ice Differential Pressure Sensor　防冰微分压力传感器

AIFS　Anti-ice Flow Sensor　防冰流量传感器

AIFSV　Anti-ice Flow Sensor Venturi　防冰流量传感器文氏管

AIPS　Anti-ice Pressure Sensor　防冰压力传感器

AIPSW　Anti-ice Pressure Switch　防冰压力开关

AIS　Anti-ice System　防冰系统

AITS　Anti-ice Temperature Sensor　防冰温度传感器

ASL　Altitude with respect to sea level　海平面高度

C c

CM　Continuous Maximum　连续最大

CV　Check Valve　单向活门

D d

DDV　Dual Distribution Valve　双路分配活门

E e

EAI　Engine Anti-ice　发动机防冰

ECDS　Eddy Current Deicing System　涡流除冰系统

EEDS　Electro-expulsive Deicing System　电动驱动式除冰系统

EIDI　Electro-impulse Deicing System　电脉冲除冰系统

EMEDS　Electro-Mechanical Expulsive De-icing System　电动机械式除冰系统

ETOPS　Extended Range Operation with Two-engine Airplanes　双发飞机扩大使用范围

F f

FIDS　Flight Ice Detection Systems　飞行结冰探测系统

FZDZ　Freezing Drizzle　冻雨

FZRA　Freezing Rain　冻雨

H h

HADV　Heated Automatic Drain Valve　自动排水加热活门

I i

ICTS　Ice-Contaminated Tailplane Stall　平尾结冰失速

ID　Icing Detector　结冰探测器

IM　Intermittent Maximum　间断最大

IMC　Instrument Meteorological Conditions　仪表气象条件

IPS　Ice Protection System　防冰系统

ISA　International Standard Atmosphere　国际标准大气

L l

LES　Large Eddy Simulation　大涡模拟

LPWS　Low Pressure Warning Switch　低压告警开关

LWC　Liquid Water Content　液态水含量

M m

MED　Median Effective Diameter　平均有效直径

MMD　Median Mass Dimension　平均重量尺寸

MSL　Mean Sea Level　平均海平面

MVD　Median Volumetric Drop　平均水滴直径

N n

NAI　Nacelle Anti-icing　短舱防冰

NAIPS　Nacelle Anti-icing Pressure Sensor　短舱防冰压力传感器

NAISOV　Nacelle Anti-icing　Shut-off Valve　短舱防冰关断活门

NLF　Natural Laminar Flow　自然层流

O o

OAT　Outside Ambient Temperature　外界环境温度

P p

PDSC　Pneumatic Deicing System Controller　气动除冰系统控制器

PFIDS　Primary Flight Ice Detector System　主飞行结冰探测系统

PIIDS　Primary In-flight Ice Detector System　主飞行结冰探测系统

PRRV　Pressure Regulating and Relief Valve　压力调节释压活门

Psia　Pounds per Square Inch Absolute　绝对压力(镑/英寸)

Psid　Pounds per Square Inch Differential　差压(镑/英寸)

Psig　Pounds per Square Inch gauge　表压(镑/英寸)

PT　Pressure Transducer　压力换能器

PWM　Pulse Width Modulation　脉冲调制解调器

S s

SLD　Supercooled Large Drop　过冷大水滴

SOV　Shut-Off Valve

T t

TAT　Total Air Temperature　总温

TAI　Thermal Anti-ice　热防冰

W w

WA Wiper Arm 雨刷臂

WAI Wing Anti-ice 机翼防冰

WAIS Wing Anti-ice System 机翼防冰系统

WAIV Wing Anti-ice Valve 机翼防冰活门

WB Wiper Blade 雨刷片

WECU Wiper Electronic Control Unit 雨刷电子控制单元

WIPC Windshield Ice Protection Controller 风挡防冰控制器

WMR Wiper Motor-reducer 雨刷减速器

WTAI Wing Thermal Anti-ice 机翼热防冰

WWS Windshield Wiper Equipment 风挡雨刷设备

T t

TAS True Airspeed 真空速

专业术语

𝒝 𝒷

保热器　heat retainer

暴雨条件　hard rain condition

备用传感器　spare sensor

标准的自然结冰条件　measured natural icing condition

标准受控结冰云团　a measured controlled simulated icing cloud

标准振荡器　reference oscillator

表面湿润系数　coefficient of surface wetting

冰雹　hail

冰风洞试验　icing wind tunnel test

冰积累　ice collection

冰脊　ice ridge

冰结在　ice build-up on the…

冰晶　ice crystal

冰桥　a bridge of ice

冰脱落周期　ice shed cycles

冰型　ice shape

波纹管　bellow

波纹管补偿器　bellows compensator

不对称结冰　asymmetric icing

不结冰状态飞行　non-icing flight

𝒞 𝒸

残留冰　residual ice

层云　stratiform clouds

持续的功率丧失　sustained power loss

持续的膨胀/收缩循环　sequential inflation/manifold

冲击水　impinging water

充气除冰带　inflatable boot

冲洗泵　washer pump

冲洗剂容器　washer reservoir

重复结冰　refreeze

抽吸　suck

抽吸模式　suction mode

除冰　de-icing

除冰定时机构　deicing timer

除冰加热器　deicing heater

除冰器(设备)　deicer

除冰气囊　deicing gasbag

除雾系统　defog system

除雨系统　rain removal system

除冰液　deicing fluid

除雨开关　rain removal switch

除冰周期　a deicing cycle

储液箱(液体防冰)　tank

穿越结冰条件　penetration of icing conditions

传感器　sensor

传感器接头　sensor terminal

垂直安定面　vertical stabilizer

D d

D 环防冰腔　D-ring spray manifold

大雨条件　heavy rain condition

带式加热器　heater tapes

导电膜　conductive coating

导电膜加温除冰　electric film heating deicing

导管故障灯　duct fail light
低速　low speed
低压开关　low pressure switch，LP
笛形管　piccolo tube
地面结冰　ground ice
地面雾　ground fog
电磁活门　solenoid valve
电动机械式除冰　electro－mechanical deicing
电机-减速器　motor-reducer
电加热探头　electrically heated probes
电加热元件　electric heating element
电加温除冰器　electro-thermal deicer
电加温导电层　conductive layer for electrical heating
电脉冲除冰　electro－pulse deicing
电脉冲除冰系统　electric impulse deicing system/electropulse deicing system
电驱动风挡雨刷　electrical windshield wipers
电热除冰　electro－thermal deicing
电热防冰系统　electro－thermal ice protection system
电热周期除冰　electro-thermal cyclic deicing
电热周期除冰系统　electro-thermal cyclic deicing system
电容器　capacitor
垫式加温器　gasket heater
电线　wire harness
电阻加温除冰器　resistive deicer
冻结分数　freezing fraction
冻雾　freezing fog
冻雨　freezing rain
短舱防冰开关　nacelle anti-ice switch
冻雨量　freezing precipitation
多孔板(机翼上的)　porous panels (on wing)

多孔除冰器　porous deicer

F f

发动机防冰　engine anti-icing

发动机进气道防冰活门　engine inlet anti-ice valve

发动机进气道热气防冰系统　engine inlet thermal anti-icing

发动机进气口除冰　engine air intake deicing

发动机前罩防冰　engine nose cowling anti-ice

发动机探头防冰　engine probe anti-ice

发动机引气活门　engine bleed air valve

发动机整流罩过热开关　engine cowl overheat switch

防冰/防雨/防雾及防霜　ice/rain/fog and frost protection

防冰和防雨系统　ice and rain protection system

防冰空气质量流量　anti-icing mass flow rate

防冰控制活门　anti-ice control valve

防冰控制开关　anti-ice control switch

防冰控制器　anti-ice controller

防冰能力　anti-icing capability

防冰设备　ice protection equipment

防冰条款　ice protect requirements

防冰系统　anti-icing system

防冰系统的延迟打开　a delay in the activation of the ice protection system

防冰系统的热能利用率　the thermal energy available for the ice protection system

防冰系统防护区　IPS coverage

防冰系统试验　test of ice protection system

防冰装置开关　deicer switch

防腐蚀蒙皮/电磁驱除式除冰　erosion skin

防腐蚀外壳/电磁驱除式除冰　erosion shield

防护表面　protected surface

防雾系统　anti-fog system
防雨剂喷嘴　rain repellent nozzle
防止静电积聚　prevention of accumulation of electrostatic charge
非对称冰脱落　asymmetric shedding of the ice
非防护表面　unprotected surface
飞机防冰系统　aircraft ice protection system
飞机防雨系统　aircraft rain protection system
飞机积冰　ice accretion of aircraft icing
飞机结冰　aircraft icing
飞溅　slush
非结冰条件　non-icing condition
飞行结冰　in-flight icing
废水排放管加热器　waste water drain mast heater
分界窄条　parting strip
分配单元　proportioning unit
分配环　distributor ring
风挡　windshield
风挡玻璃窗加热元件　window heat components
风挡玻璃除冰　windshield deicing
风挡除雨功能　windshield rain removal function
风挡除雨系统　windshield wiper system
风挡防冰　windshield ice protection
风挡防冰开关　windshield anti-icing switch
风挡防雾风扇　windshield defogging fan
风挡防雨　windshield rain protection
风挡加温控制单元　window heat control units
风挡加温控制器　windshield heat controller
风挡排雨剂　windshield rain repellent
风挡喷射管　windshield spray bay
风挡曲率　windshield curvature
风速　wind speed

风挡洗涤器　windshield washer
风挡洗涤系统　windshield washer system
风挡液泵　windshield pump
风挡雨刷(刮雨器)　windshield wiper
风挡雨刷开关　windshield wiper switch
风挡雨刷控制装置　windshield wiper control unit
风挡雨刷转弯器　windshield wiper converter
缝合紧密　stitching tightness

G g

干防冰　dry anti-icing
干结冰　dry icing
干空气飞行试验　dry air flight test
干湿表面温差　temperature difference between dry surface and wet surface
干态　run dry
干燥空气环境地面试验　dry air ground test
高速　high speed
高压开关　high pressure switch
告警灯　warning light
攻角叶片加温器　angle of attack/AOA vane heater
功率损失不稳定性　power loss instability
供水管加温　water supply lines heat
刮刷角度　wiping angle
刮刷区域　sweeping area/swept area
刮刷速度　wiping speed
刮刷循环　wiping stroke
刮水角　wiping angle
关键表面　critical surface
关键冰型　critical ice shape
关键飞行冰型　critical in-flight ice shape

关键结冰防护设计点 the critical ice protection design points

过冷大水滴 supercooled large drops

过冷水滴 supercooled water droplet

过滤装置 filter assembly

过热传感器 overheat sensor

过热恒温器 overheat thermostats

H h

海里 nautical miles

恒温器 thermostat

恒温调节加温器 thermostatically controlled heater

后部结冰形成的界限 the rear ice formation limit

环形加热器 annular heater

回流 reversal airflow

汇流条 bus

混合冰 mixed ice

混合结冰条件 mixed conditions

混频器 mixer

J j

积冰强度/速率 intensity of icing/icing rate

积冰模拟试验 test with simulation of icing

激光水滴测量记录仪 a laser-based droplet measuring and recording instrument

积云 cumuliform clouds

机体除冰 airframe deicing

机械除冰系统 mechanical deicing system

机械除冰设备 mechanical deicer

机翼除冰 wing deicing

机翼防冰灯 wing icing detection lights

机翼防冰活门 wing anti-ice valve

机翼防冰管路　wing anti-icing ducts

机翼防冰控制开关　wing anti-ice control switch

机翼防冰控制开关与继电器　wing anti-ice control switch and relays

机翼气囊　wing boots

机翼前缘防冰　wing leading edge anti-ice

机翼热力防冰　wing thermal anti-ice

机翼与发动机防冰控制面板　wing and engine anti-icing control panel

计量泵　metering pump

计算冰型　computed ice

驾驶舱窗防冰　flight compartment window anti-ice

驾驶舱视界　pilot compartment view

驾驶舱主风挡　main cockpit windshields

驾驶舱最小外部视界　pilot compartment minimum external view

驾驶员视界　vision for pilot and copilot

驾驶员调节器　pilot regulator

加热功率　heated power

加热控制器　heat controller

加热器　heater strips

加温器接头　heater terminal

加热器开关灯　heater switch light

加热条带　heating braid

加热元件　heating element

加速老化　an accelerated ageing

加温除冰器　thermal deicer

间断速度　intermittent speed

间断最大结冰条件　the intermittent maximum

剪切力　shear

桨毂整流罩加温　hub heating

降雨条件　precipitation conditions

降雨速率　precipitation rate

胶条　rubber

结冰　ice accretion/accretion of ice
结冰包线　the icing envelop
结冰传感器/探测器　ice detector
结冰传感器受感器　pick-up unit of icing detector
结冰传感器随动器　follow-up unit of icing detector
结冰防护系统能量消耗　the power extraction associated with ice protection system operation
结冰飞行试验　icing flight test
结冰风洞　ice wind tunnel
结冰极限　ice accretion limit/icing extent
结冰极限飞行速度　limiting flight velocity for icing
结冰计算状态　icing calculation condition
结冰模拟　ice accretion simulation
结冰喷水飞机(洒水机)　an icing tanker
结冰气象　icing meteorology
结冰气象参数　meteorological parameter of icing
结冰强度　icing intensity
结冰认知方法　the icing recognition means
结冰设计标准　icing design criterion
结冰适航包线　icing certification envelop
结冰适航要求　icing certification requirements
结冰速率　the rate of ice buildup
结冰探测灯　ice detection/inspection light
结冰探测系统　ice detection system
结冰探针　an ice evidence probe
结冰条件的形成　the onset of icing conditions
结冰脱落　the removal of ice
结冰污染物　icing contamination
结冰相关性能损失　icing-related performance losses
结冰信号传感器　ice-detector probe
结冰异常现象　icing anomalies

结冰云　icing cloud
结冰指示器　ice indicator
进气道除冰　inlet/intake deicing
进气道空气加热器　intake air heater
进气道气囊　air intake boots
进气系统结冰防护　induction system icing protection
静压管加温器　static heater
局部水收集系数　local water catch efficiency

K　k

空气数据传感器加温板　air data sensor heat panel
空气数据传感器加温系统　air data sensor heat system
空速管/动压管加温器　pitot heater
控制恒温器　control thermostat
控制器/电脉冲除冰　program switching
扩散缝口　diffuser slot

L　l

拉格朗日水滴轨迹计算　Lagrangian drop or detected ice accretion
朗缪尔分布　Langmuir distribution
老化　aging
连续最大结冰条件　the continuous maximum ice-formation condition
流湿防冰　raining wet anti-icing/imperfect evaporation anti-icing
露点温度　dew-point temperature
流体流动路线　fluid flow paths
螺旋桨防冰　propeller anti-ice
螺旋桨除冰　propeller deicing
螺旋桨加温控制器　propeller heat control unit
螺旋桨电加热除冰带　electrically heated propeller boot
氯丁橡胶　neoprene

M m

矛状冰 spearhead ice
脉动除冰带 pulsating deicer boot
明冰 glaze ice/clear ice
模拟冰型 simulated ice shapes
模拟结冰 simulated icing
模拟结冰飞行试验 simulation icing flight test
模拟结冰手段 simulated icing tool
模型阻塞比 model blockage
磨耗损失 abrasion loss

N n

NACA 进气道 NACA inlet
黏结 bond/adhesion
黏结剂 glue
内部泄漏 internal leakage

P p

排水管加温 drain line heat
排水装置加热器 drain mast heater
排雨刷 rain repellent
排雨剂存储器 rain repellent bottle
排雨剂罐 repellent canister
喷沙 sandblast
喷雾嘴 spray nozzle
膨胀/收缩速率 inflation/deflation speed
膨胀管除冰 inflatable tube deicing

Q q

气动脉冲除冰 pneumatic impulse deicing

气动带除冰系统　pneumatic boot deicing system/inflatable tube deicing system

气动带除冰　pneumatic boot deicing

气动除冰带　pneumatic deicer boot

气动前缘除冰带　pneumatic leading edge boot

气动除冰系统　pneumatic boot deicing system/inflatable tube deicing system

气动安全活门/减压活门　pneumatic relief valve

气动除冰系统　pneumatic deicing system

汽化器除冰器　carburetor deicer

气囊的膨胀与收缩　inflation or deflation boots

气囊总容积　total boots volume

气囊移除与操作时间　boot removals and boots operating hours

气热防冰系统　hot air ice protection system/hot air anti-ice system

气热除冰　hot air deicing

前整流罩压力开关　nose cowl pressure switch

潜在的结冰条件　potential icing conditions

侵蚀　erosion

清洁航空器　clear aircraft

清除结冰　clear of ice

轻微积冰　light rime icing

球形温度补偿器　spherical temperature compensator

全静压管/空速管除冰　pitot deicing

全静压探头防冰　pitot-static probe anti-ice

全静压探头加温控制器　pitot-static probe heat control unit

全翼展冰型　full-span ice shape

R r

燃油/滑油加温器　fuel/oil heater

热刀　heat knife

热防冰　thermal anti-icing

热风加热器　storage heater

热空气除冰设备　hot-air deicer

热空气热除冰前缘系统　hot air thermal deice leading edge system

热空气热防冰前缘系统　hot air thermal anti-ice leading edge system

热力防冰系统　thermal anti-icing system/thermal ice protection system

热敏电阻　thermistor

热能防冰系统热量包络图　a thermal profile of an operating thermal ice protection system

热气表面加热器　preheat-air surface heater

热线风速仪(用来测量液态水含量)　a hot-wire anemometer-based instrument

人造冰　artificial ice

任务剖面　mission profiles

容积平均水滴直径　volume median droplet diameter

柔性转接性　flexible coupling

S s

三相负载　a three phase load

丧失除冰功能　total loss of deicing function

上下面后部水滴冲击界限　upper and lower aft droplet impingement limits

上翼面防护区域　ice protection area for upper wing surface

射流　jet flow

升华结冰　sublimation icing

失效冰型　failure ice

湿防冰　wet anti-icing

湿态防冰　running-wet

湿态防冰系统　running-wet ice protection system

使用速度包线　operational speed envelope

释压　relief pressure

收集区　trapping zone

收集系数　collection efficiency
收缩套管　telescoping duct
手持电筒　handheld flashlight
手动超控　manual override
输出轴　output shaft
双角状冰　double-horn ice
霜冰　rime ice
水滴尺寸　droplet size/drop size
水滴滑动装置(可用来测量水滴直径)　a drop slide device
水滴积冰　waterdrop icing
水滴浓度　water drop concentration
水滴收集　water catch
水滴收集参数　droplet trapping parameter
水滴直径　waterdrop diameter
水滴撞击极限　droplet impingement limits
水滴撞击特性　droplet impingement characteristic
水管及废水管防冰　water and waste lines anti-ice
水膜　water film
水平安定面　horizontal stabilizer
水收集系数　water catch efficiency
水箱加温器　water-tank heater
45 分钟待机条件　45-minute hold condition
速度包线　speed envelope
算术平均水滴直径　mean arithmetic droplet diameter

T　t

探冰棒/头　probe icing detection bar
探测头加温器　probe heater
提前结冰　preactivation ice
调节关断活门　regulator shut-off valve

调压器　voltage regulator

条纹板防冰　strake anti-ice

停放角度　parking angle/park angle

停放位置　parking position

吞冰　ice ingestion

脱落　shedding

脱落冰轨迹　shed-ice trajectories

脱落冰危害　shed-ice hazards

W　w

尾翼除冰　tail deicing

位置开关　position switch

温度补偿器　temperature compensator

温度-高度包线　the temperature-altitude envelope

温度控制器　temperature control unit

稳定结冰　stabilized ice accretion

稳定性操作　stabilized operation

无机玻璃　inorganic glass

无结冰防护的表面　surfaces without ice protection

X　x

细雨　freezing drizzle

下翼面防护区域　ice protection area for lower wing surface

弦向加热区　chordwise parting strip

限流器　restrictor

线圈　coil

橡胶管　rubber tubes

橡皮除冰带　rubber-boot deicer

斜坡式/比例加温　ramp warmup

信号灯　annunciator light

信号装置　annunciation

形状记忆合金除冰系统　shape memory alloy de-icing system
雪　snow

Y y

压降　pressure drop
压力表　pressure gauge
压力开关　pressure switch
严酷冰型　critical ice shape
严酷结冰/严重积冰　severe icing
严重丧失功率或推力　serious loss of power or thrust
液态含水量　liquid water content
液体除冰　fluid deicing
液体除冰设备　deicing liquid devices
液体除冰系统　fluid deicing system
液体除雨剂　liquid repellent solution
液体防冰系统　fluid anti-icing system/fluid ice protection system
液压刮雨器　hydraulic windshield wiper
已知的或者观察到的或者探测到的结冰　known or observed or detected ice accretion
已知的结冰条件　known icing conditions
易燃流体　flammable fluid
溢流冰　runback ice
溢流水　runback water
翼肋缘条　rib flange
阴影区　shadowed zone
迎角传感器防冰　angle of attack probe/AOA anti-ice
引气供给管　bleed air supply duct
引射管　ejector tube
饮用水加热器　potable water heater
有机玻璃　plexiglas
有效平均水滴直径　mean effective droplet diameter

雨刷臂　wiper arm
雨刷刮板　wiper blade
雨刷控制　wiper control
雨刷马达转向器装置　motor converter unit
雨中结冰　icing in precipitation
预测结冰条件　forecast icing conditions
预测人造冰型　predicted artificial ice shape
预紧力　pretightening force
云中结冰　icing in cloud

Z　z

遭遇结冰　an icing encounter
展向加热区　spanwise heating zone
真空引射器　vacuum ejector
蒸发防冰　evaporative anti-icing
正常传感器　normal sensor
织物延伸率　fabrics elongation
执行器/电磁驱除式除冰　actuator
执行器外壳/电磁驱除式除冰　actuator housing
执行器支架安装横梁　actuator tray mount rail
执行器装配支架　actuator assembly carrier tray
直流无刷电机　DC brushless motor
中度积冰　moderate icing
中度结冰　moderate ice
重度结冰　heavy icing
重要视野区域　important field of vision
周期加热区　cycled heating zone
周期间结冰　intercycle ice
周期性防冰系统　cyclic ice protection systems
主防冰管路　main anti-ice ducts
转换器/电脉冲除冰　transformer

撞击极限　impingement limit

撞击极限加热器　bonded heater to impingement limits

自然冰型　natural ice shape

自然结冰飞行试验　natural icing flight test

自然结冰飞行试验　natural icing flight test

自然结冰条件　natural icing condition

总大气温度探头加温器　total air temp/TAT probe heater

总收集系数　total water catch efficiency

总温传感器防冰　total air temperature/TAT probe anti-ice

最大结冰防护载荷　maximum ice protection load

最关键人造冰型　the most critical artificial ice shape

左机翼前梁轴线　the axis of left wing front spar

座舱玻璃防冰　flight compartment window anti-icing

座舱视界　pilot compartment view

控制和软件

缩略语

A a

A/D Analog-To-Digital 模数转换

ACT Active 激活

ADA Computer Programming Language 计算机编程语言

ARINC Aeronautical Radio Incorporated 航空无线电公司

AUTO Automatic 自动

AVAIL Available 可用的

B b

BIT Built-in Test 自检测

BITE Built-in Test Equipment 自检测设备

BNR Binary 二进制

BSP Board Support Package 板级支持包

C c

CAN Controller Area Network 控制器局域网

CAR Cause Analysis and Resolution 原因分析和决定

CAS Crew Alerting System 机组提示系统

CASE Computer Aided Software Engineering 计算机辅助软件工程

CB Circuit Breaker 跳开关/断路器

CBIT Continuous BIT 连续自检测

CCB Configuration Control Board 配置控制委员会

CDR Critical Design Review 关键设计评审

CFD Control Flow Diagram 控制流图

CM Configuration Management 配置管理/构型管理

COM Computer Operation Manual 计算机操作手册

COTS Commercial Off The Shelf 商业货架产品

CPD Circuit Protection Device 电路保护设备

CPM Computer Programming Manual 计算机编程手册

CR Change Request 更改请求

CRB Change Review Board 更改评审委员会

CRC Cyclic Redundancy Check 循环冗余校验

CSA Code Safety Analysis 代码安全性分析

CSC Computer Software Component 计算机软件部件

CSCI Computer Software Configuration Item 计算机软件配置项

CSU Computer Software Unit 计算机软件单元

CTL/CTRL Control 控制

CU Control Unit 控制装置

CV Check Valve 单向活门

D d

DAL Design Assurance Level 设计保证等级

DAR Decision Analysis and Resolution 决策分析和决定

DBDD Database Design Description 数据库设计说明

DC Direct Current 直流电

DDP Declaration of Design and Performance 设计性能声明

DDV Dual Distributing Valve 双重分配活门

DER Designated Engineering Representative 工程委任代表

DFD Data Flow Diagram 数据流图

DICD Digital Interface Control Document 数字接口控制文件

DID Data Item Description 数据项描述

DISC Discrete 离散信号

DITS Digital Information Transfer System 数字信息传递系统

DLA Design Logic Analysis 设计逻辑分析

E e

EICAS　Engine Indicating and Crew Alerting System　发动机指示和机组提示系统

EMI　Electromagnetic Interference　电磁干扰

EPS　Electrical Power System　电源系统

F f

FCA　Functional Configuration Audit　功能性配置审计

FCU　Flush Control Unit　冲水控制装置

FDIR　Fault Detection, Isolation, and Recovery　故障探测、隔离和恢复

FDR　Flight Data Recorder　飞行数据记录仪

FHA　Failure Hazard Assessment　失效危险评估

FPM　Feet Per Minute　英尺/分钟

FR　Failure Rate　失效率

FWD　Forward　前

G g

GG　Generic Goal　共用目标

GND　Ground　地

GP　Generic Practice　共用实践

GPM　Gallon Per Minute　加仑/分钟

GSE　Ground Support Equipment　地面支持设备

H h

HADV　Heated Automatic Drain Valve　加热自动排水活门

HCI　Hardware Configuration Item　硬件配置项

HCI　Human-Computer Interface　人机接口

HIRF　High Intensity Radiated Fields　高强度辐射场

HLR　High Level Requirement　顶层需求

HP High Pressure 高压

I i

IA Independent Assessment 独立评估

IBIT Initiated BIT 启动自检测

ICD Interface Control Document 接口控制文件

IDD Interface Design Description 接口设计说明

IECS Integrated Electro-mechanical Control System 综合机电控制系统

IEEE Institute of Electrical and Electronics Engineers 电气和电子工程师协会

INFO Information 信息

INOP Inoperative 不工作

IOS International Organization for Standardization 标准化国际组织

IPM Integrated Project Management 集成的项目管理

IRS Interface Requirement Specification 接口需求规格说明

ISOV Isolator Shut-Off Valve 隔离关断活门

IV&V Independent Verification and Validation 独立验证和确认

J j

JCDP Joint Conceptual Definition Phase 联合概念定义阶段

JDP Joint Definition Phase 联合定义阶段

L l

LAI Left Air Intake 左发入口

LED Light Emitting Diode 发光二极管

LH Left Hand 左侧

LLR Low Level Requirement 底层需求

LP Low Pressure 低压

LPWS Low Pressure Warning Switch 低压告警开关

LRU　Line Replaceable Unit　航线可替换单元

LW　Left Wing　左翼

M　m

MA　Measurement and Analysis　测量与分析

MBIT　Maintenance BIT　维护自检测

MIL-STD　Military Standard　军用标准

MIN　Minimum　最小

ML　Maturity Level　成熟度等级

MNWF　Must Not Work Function　非必需的工作功能

MTBUR　Mean Time Between Unscheduled Repair and Removals　非计划维修和替换平均时间

MVCP　Module Verification Cases and Procedures　模块验证用例和程序

MVR　Module Verification Report　模块验证报告

MWF　Must Work Function　必需的工作功能

N　n

N/A　Not Applicable　不适用

NCD　Nolinear Control Design　非线性控制设计

NDI　Non-Developmental Item　非开发项

NDS　Non-Developmental Software　非开发软件

NORM　Normal　正常

NVM　Non-Volatile Memory　非易失存储器

O　o

OAT　Outside Air Temperature　外界大气温度

OCD　Operational Concept Description　运行方案说明

OID　Organizational Innovation and Deployment　组织创新和部署

OMS　Onboard Maintenance System　机载维护系统

OO　Object Oriented　面向对象

OOA Object Oriented Analysis 面向对象的分析

OOD Object Oriented Development 面向对象的开发

OPD Organizational Process Definition 组织过程定义

OPF Organizational Process Focus 组织过程焦点

OPP Organizational Process Performance 组织过程绩效

OT Organizational Training 组织培训

OTS Off the Shelf 货架

OVHD Overhead 顶部

OVRD Override 超控

P p

P&T Pressure and Temperature 压力和温度

PA Process Area 过程域

PBIT Power-on BIT 加电自检测

PCA Physical Configuration Audit 物理配置审计

PDL Program Description Language 程序描述语言

PDR Preliminary Design Review 初步设计评审

PDS Previously Developed Software 以前开发软件

PFD Primary Flight Display 主飞行显示器

PHA Preliminary Hazard Analysis 初步危险性分析

PI Product Integration 产品集成

PID Proportional Integral Differential 比例积分微分

PLC Programmable Logic Controller 可编程逻辑控制器

PLD Programmable Logic Device 可编程逻辑器件

PMC Project Monitoring and Control 项目监控

POST Power On Self Test 加电自测试

PP Project Planning 项目策划

PPQA Process and Product Quality Assurance 过程和产品质量保证

PR Problem Report 问题报告

PRRV Pressure Regulating and Relief Valve 压力调节关断活门

PRSOV Pressure Regulating And Shutoff Valve 压力调节关断

活门

PSAC Plan for Software Aspects of Certification 软件审查计划

PT Pressure Transducer 压力变换器

PWM Pulse-Width Modulation 脉宽调制

Q q

QA Quality Assurance 质量保证

QMTC Qualified Model Test Coverage Tool 合格模型测试覆盖率工具

QPM Quantitative Project Management 定量项目管理

QTE Qualified Test Environment 合格测试设备

QTY Quality 质量

R r

RAI Right Air Intake 右发入口

RAM Random Access Memory 随机存取存储器

RD Requirements Development 需求开发

REF Reference 参考/基准

REQM Requirements Management 需求管理

RH Right Hand 右侧

ROM Read Only Memory 只读存储器

RSKM Risk Management 风险管理

RTCA Radio Technical Commission for Aeronautics 〈美〉航空无线电技术委员会

RTOS Real Time Operating System 实时操作系统

RVCP Requirements Verification Cases and Procedures 需求验证用例和程序

RVDT Rotary Variable Differential Transducer 旋转可变差动传感器

RVR Requirements Verification Report 需求验证报告

RW Right Wing 右翼

S s

S&MA　Safety and Mission Assurance　安全和任务保证

SA　Software Assurance　软件保证

SADA　Safety Architectural Design Analysis　安全架构设计分析

SAM　Supplier Agreement Management　供方协议管理

SAS　Software Accomplishment Summary　软件设计总结

SAT　Static Air Temperature　大气静温

SCCSC　Safety-critical Computer Software Component　安全关键的计算机软件部件

SCI　Software Configuration Index　软件配置清单

SCM　Software configuration management　软件配置管理

SCMP　Software Configuration Management Plan　软件配置管理计划

SCMR　Software Configuration Management Report　软件配置管理报告

SCS　Software Code Standard　软件编码标准

SCSR　Software Configuration Status Report　软件配置状态报告

SDD　Software Design Description　软件设计说明

SDDA　Safety Detailed Design Analysis　安全性详细设计分析

SDP　Software Development Plan　软件开发计划

SDS　Software Design Standard　软件设计标准

SDSR　Software Development Summary Report　软件研制总结报告

SDTD　Software Development Task Description　软件研制任务书

SECI　Software Life Cycle Environment Configuration Index　软件生存周期环境配置列表

SEE　Software Engineering Environment　软件生存周期环境

SEU　Single Event Upset　单事件反转

SFMEA　Software Failure Modes and Effects Analysis　软件失效模式和影响分析

SFTA　Software Fault Tree Analysis　软件故障树分析

SG　Specific Goal　专用目标
SIOM　Software Input Output Manual　软件输入/输出手册
SIP　Software Installation Plan　软件安装计划
SIS　Software Interface Specification　软件接口规范
SOI　Stage of Involvement　介入阶段
SOW　Statement of Work　工作说明
SP　Specific Practice　专用实践
SPA　Software Product Assurance　软件产品保证
SPDT　Single Pole Double Throw　单刀双掷
SPDU　Secondary Power Distribution Unit　二次配电单元
SPS　Software Product Specification　软件产品规格说明
SQA　Software Quality Assurance　软件质量保证
SQAP　Software Quality Assurance Plan　软件质量保证计划
SQAR　Software Quality Assurance Report　软件质量保证报告
SRD　Software Requirements Document　软件需求文档
SRR　Software Requirements Review　软件需求评审
SRS　Software Requirements Specifications　软件需求规格说明
SRS　Software Requirements Standard　软件需求标准
SSAR　Software Safety Analysis Report　软件安全性分析报告
SSHA　Software Subsystem Hazard Analysis　软件子系统危险性分析
SSM　Sign Status Matrix　符号状态矩阵
SSPC　Solid-State Power Controller　固态功率控制器
SSR　Software Specification Review　软件规范评审
SSRA　Software Safety Requirements Analysis　软件安全性需求分析
STAT　Status　状态
STBY　Standby　待机
STD　Software Test Description　软件测试说明
STD　Standard　标准
STP　Software Test Plan　软件测试计划
STR　Software Test Report　软件测试报告

SUM　Software User Manual　软件用户手册

SVD　Software Version Description　软件版本说明

SVP　Software Verification Plan　软件验证计划

T　t

TB　Terminal Board　接线端子板

TEMP　Temperature　温度

TRD　Technical Requirements Document　技术要求文件

TRR　Test Readiness Review　试前评审

TS　Technical solution　技术解决方案

TSO　Technical Standard Order　技术标准说明

U　u

UML　Unified Modeling Language　统一建模语言

V　v

VAC　AC Voltage　交流电压

VAL　Validation　确认

VDC　DC Voltage　直流电压

Ver　Verification　验证

VSO　Virtual System Operation　虚拟系统运行

W　w

WBS　Work Breakdown Structure　工作分解结构

WIPS　Wing Ice Protection System　机翼结冰保护系统

WLS　Water Level Sensor　清水液位传感器

WP　Work Product　工作产品

WWSC　Water Waste System Controller　污水系统控制器

专业术语

A a

安全释压活门　safety relief valve

安全性/保密性　security

B b

版本　version

比较器　comparator

编程/程序设计　programming

编程语言/程序设计语言　programming language

编译　compile

编译程序/编译器　compiler

变流器　inverter

标识符　identifier

标志　flag

标准实施器　standards enforcer

别名　alias

不工作　inoperative

不工作的　INOP

不完全的隐错排除　imperfect debugging

部件/组成部分　component

部件号　part number

C c

参数　parameter

参数整定　parameter setting

操作符　operator

测试　testing
测试报告　test report
测试范围　test coverage
测试规程　test procedure
测试计划　test plan
测试可重现性　test repeatability
测试驱动程序　test driver
测试日志　test log
测试数据　test data
测试数据生成器　test data generator
测试台　test bed
测试用例　test case
测试用例生成器　test case generator
测试有效性　test validity
层次结构　hierarchy
层次结构分解　hierarchical decomposition
插头　receptacle plug
插座　socket
插座　receptacle
产品　product
产品库　product library
超控　override
程序保护　program protection
程序变异　program mutation
程序规格说明　program specification
程序库　program library
程序块　program block
程序扩展　program extension
程序设计支持环境　programming support environment
程序探测　program instrumentation
程序体系结构　program architecture

程序支持库　program support library
程序综合　program synthesis
出错分析　error analysis
出错类别　error category
出错预测　error prediction
出口　exit
处理　process
传感器　transducer/sensor
错误/出错/误差　error

D　d

搭接片　bonding jumper
大容量存储器　mass memory
代码/编码　code
代码审查　code inspection
代码审计　code audit
代码生成器　code generator
代码走查　code walk-through
带保护罩的开关　guarded switch
带故障工作　fail operational
单向活门　non-return valve
导线　wire
低级需求　low-level requirements
地线故障　ground fault
地址　address
地址空间　address space
递归例行程序　recursive routine
电磁活门　solenoid controlled valve
电磁线圈　solenoid
电气接口控制文件　electrical interface control document
电熔丝　electrical fuse

迭代　iteration
独立性　independence
独立验证和确认　independent verification and validation
短舱　nacelle
短路　short circuit
断开　de-energize
队列　queue
多版本非相似软件　multiple-version dissimilar software

E　e

二进制编码的十进制　binary coded decimal

F　f

发光二极管　light emitting diode
发行　release
翻译程序/转换程序　translator
反常行为　anomalous behavior
反馈　feedback
反逻辑　negative logic
反面　offside
飞行阶段　phase of flight
非门　not gate
非线性过程控制 nonlinear process control
非线性 PID　nonlinear PID
分时　time sharing
分析模型　analytical model
封装　encapsulation
符合性方法　means of compliance
负释压活门　negative relief valve
复杂性　complexity
副作用　side effect

赋值语句　assignment statement

覆盖　overlay

覆盖范围分析　coverage analysis

G g

改正性维护　corrective maintenance

高级需求　high-level requirements

高级语言　high level language/higher order language

隔离活门　isolation valve

工具鉴定　tool qualification

公差　tolerance

功能　function

功能部件　functional unit

功能分解　functional decomposition

功能规格说明　functional specification

功能性配置审计　functional configuration audit

功能需求　functional requirement

供方　supplier

固件　firmware

固态电门　solid state switch

故障　fault

故障旗　fault flag

故障容差　fault tolerance

故障消极防护　fail passive

关键部分优先　critical piece first

关键段/临界段　critical section

管理程序　supervisory program

规格说明/规范　specification

规格说明语言　specification language

规模估计　sizing

国际标准大气 international standard atmosphere，ISA

H h

函数 function

航空电子学 avionics

毫巴 millibar

合格性测试 qualification testing

合同 contract

恒温电门 thermostatic switch

恒温器 thermostat

宏 macro

宏指令 macro instruction

后备/后援 back up

环境压力 ambient pressure

回测/回绕 wrap around

回归测试 regression testing

汇编 assemble

汇编语言 assembly language

汇流条/总线 bus

活动 activity

活动文件 active file

或非门 nor gate

或门 or gate

获取 acquisition

J j

机场标高 airport elevation/field elevation

机器语言 machine language

基线 baseline

集成 integration

集成测试 integration testing

集成电路　integrated circuit

计时分析程序　timing analyzer

计算机程序　computer program

计算机程序注释　computer program annotation

计算机软件配置项 computer software configuration item

记录　record

继电器　relay

兼容性　compatibility

监护　confinement

健壮性　robustness

鉴定　qualification

鉴定需求　qualification requirement

交付　delivery

交替　alternate

接口控制文件　interface control document，ICD

接口需求　interface requirement

结构化程序　structured program

结构化程序设计　structured programming

结构化程序设计语言　structured programming language

结构化设计　structured design

解除预位　disarm

解释　interpret

解释程序　interpreter

解释器　interpreter

紧固件　fastener

进程　process

精度　precision

静态分析　static analysis

静态分析程序　static analyzer

静态分析器　static analyzer

静态结合　static binding

决策　decision
决策范围　decision coverage

K　k

开发方法学　development methodology
开放系统　open system
开关　switch
开路　open circuit
可测试性　testability
可接近性　accessibility
可靠性　reliability
可靠性模型　reliability model
可靠性评估　reliability assessment
可靠性评价　reliability evaluation
可靠性数据　reliability data
可靠性增长　reliability growth
可扩展性　scalability
可维护性　maintainability
可移植性　portability
可用性　availability
可用性模型　availability model
可重复性　repeatability
可重用性/复用率　reusability
可追踪性　traceability
空地传感器　air/ground sensor
控制结构　control structure
控制开关　control switch
控制流　flow of control
控制律　control law
控制数据　control data
控制语句　control statement

块结构语言 block structured language
框图 block diagram

L l

里程碑 milestone
利益相关方 stakeholder
例程 routine
连接 connection
临界的/关键的 critical
流程图 flowchart
鲁棒性 robustness
路径表达式 path expression
路径分析 path analysis
路径条件 path condition
逻辑电路示意图 logic diagram
逻辑门 logic gate

M m

命令语言 command language
模糊 PID fuzzy PID
模糊控制 fuzzy control
模拟 simulation
模拟器 simulator
目标参数 target parameter
目标程序 object program
目标码 object code
目标语言 target language
目标机 target machine

N n

内核 kernel
内聚度 cohesion

O o

耦合度 coupling

P p

排气活门 outflow valve
派生需求 derived requirements
判定表 decision table
配置 configuration
配置标识 configuration identification
配置管理 configuration management
配置控制 configuration control
配置控制委员会 configuration control board
配置审计 configuration audit
配置项 configuration item
配置项标识 configuration identification
配置状态报告 configuration status accounting
评价 evaluation
评审 review

Q q

嵌入式计算机系统 embedded computer system
嵌入式软件 embedded software
驱动程序 driver
确认 validation

R

热电偶　thermocouple
热电瓶汇流条　hot battery bus
热敏电阻　thermistor
人工语言　artificial language
人机接口　man-machine interface
任务　task
容错　fault tolerance/tolerance
冗余　redundancy
软件　software
软件产品　software product
软件单元　software unit
软件档案库　software repository
软件更改　software change
软件工程　software engineering
软件工具　software tool
软件监督程序　software monitor
软件接口　software interface
软件结构　software architecture
软件经验数据　software experience data
软件开发周期　software development cycle
软件可靠性　software reliability
软件库　software library
软件库管理员　software librarian
软件潜行分析　software sneak analysis
软件生存周期　software life cycle
软件数据库　software data base
软件维护　software maintenance
软件需求　software requirement
软件硬件接口　software-hardware interface

软件综合　software integration
软硬件综合　hardware/software integration

S s

删除　del
商业货架产品软件　commercial off-the-shelf software
设计　design
设计方法学　design methodology
设计规格说明　design specification
设计阶段　design phase
设计评审　design review
设计需求　design requirement
设计语言　design language
神经 PID　neural-PID
审计　audit
失效　failure
失效恢复　failure recovery
失效率　failure rate
失压　decompression
实参　actual parameter
实时　real time
实现　implementation
实用软件　utility software
时滞系统　time-delay system
适航性　airworthiness
适应性　adaptability
适应性维护　adaptive maintenance
释压活门　relief valve
输出　output
输出断言　output assertion
树　tree

数据　data
数据结构　data structure
数据可靠性　reliability numerical
数据库/数据基　data base
数据类型　data type
数据字典　data dictionary
数据总线　data bus
双份编码　dual coding
顺序进程　sequential processes
瞬变　transient
伺服系统　servo system
死码　dead code
宿主机　host machine
宿主计算机　host computer
算法　algorithm
算法分析　algorithm analysis

T　t

弹簧加载的　spring loaded
提示　prompt
体系结构　architecture
体系结构设计　architectural design
条件/决策覆盖　condition/decision coverage
条件控制结构　conditional control structure
通电　energize
通告　announciations
通信接口　communication interface
通信协议　communication protocol
同侧　onside
同轴电缆　coaxial cable
统计测试模型　statistical test model

退出　exit

退役　retirement

吞吐量　throughput

托架　bracket to enclose or surround

W　w

外界大气温度　outside air temperature

完全正确性　total correctness

完整性　integrity

危急程度　criticality

微程序　microprogram

微动电门/微动开关　micro switch

微码　microcode

伪码　pseudo code

未激活代码　deactivated code

文氏管　venturi

稳定性　stability

物理配置审核　physical configuration audit

物理需求　physical requirement

X　x

系统测试　system testing

系统软件　system software

下推式存储器　pushdown storage

下载　download

线性可变差动变压器　linear variable differential transformer

线性 PID　linear PID

详细设计　detailed design

项目计划　project plan

效率　efficiency

协议　protocol

卸出/转储　dump
信号　signal
信号量　semaphore
形参　formal parameter
性能　performance
性能规格说明　performance specification
性能需求　performance requirement
修补　patch
修改　modification
虚拟存储器　virtual storage
虚拟机　virtual machine
需方　acquirer
需求的规格说明语言　requirements specification language
需求分析　requirements analysis
需求规格说明　requirements specification
需求阶段　requirements phase
旋转可变差动传感器　rotary variable differential transducer

Y y

压力调节关断活门　pressure regulating and shutoff valve
延时电路　time delay
严重性　severity
验收测试　acceptance testing
验收准则　acceptance criterion
验证　verification
异常　exception
遗传算法　genetic algorithm
易失存储器　volatile memory
引导程序　bootstrap
引导装入程序　bootstrap loader
引射口　ejector

应用软件　application software
硬件　hardware
硬件配置项　hardware configuration item
用户　user
用户文档　user documentation
有限状态机　finite state machine
与非门 NAND　gate
与国际标准大气之偏差　isa deviation
与门　and gate
语法　syntax
语法分析　parse
语义　semantics
预编译程序　precompiler
预处理程序　preprocessor
预位　arm
源程序　source program
源语言　source language
运行测试　operational testing
运行可靠性　operational reliability

Z z

再循环空气　recirculated air
栈　stack
招标/标书　request for proposal
诊断　diagnostic
正确性　correctness
正确性证明　correctness proof
支持软件　support software
执行　execution
执行时间　execution time
指令　instruction

指令集合　instruction set
指令集合结构　instruction set architecture
指针　pointer
质量　quality
质量度量　quality metric
中断　interrupt
终止　exit
终止性证明　termination proof
重定位机器代码　relocatable machine code
周围环境　ambient
注释　comment
转换　changeover/conversion
转换准则　transition criteria
专家 PID　experts PID
桩模块　stub
装入程序　loader
状态图　state diagram
追踪程序　tracer
准确/准确度　accuracy
子程序　subprogram
子例行程序/子例程　subroutine
字　word
字节　byte
自底向上　bottom up
自底向上设计　bottom up design
自顶向下　top-down
自顶向下测试　top-down testing
自顶向下设计　top-down design
自动测试用例生成器　automated test case generator
自动设计工具　automated design tool
自动验证工具　automated verification tools

自动验证系统　automated verification system

踪迹/追踪　trace

总管　manifold

总和校验　checksum

走查　walk-through

组织过程　organization process

应急救生系统

导 语

弹射座椅是飞机应急救生系统中重要的组成部分。

当一个飞机乘员必须紧急离开飞机时，弹射座椅应该能让他（她）安全地离开。在大多数情况下，如果一个飞行员发现自己在 20 s 内有麻烦，理应很容易地脱离座椅并且在飞机的一侧跳下，这样的话降落伞就可以安全地部署了。然而，随着飞行器动力和性能的不断提升，从飞机侧边跳下的风险性问题就迅速变得极为显著。弹射座椅对于现代飞行器来说，在紧急情况下对飞行员的安全是十分重要的。弹射座椅系统能够部署降落伞并且保障乘员安全的着陆。从飞机弹射的时间比目前乘员拉弹射手柄直到降落伞在顶上展开的时间要少 2 s。为了让弹射座椅能拯救乘员的生命，系统中近 500 个部件必须适当并且快速而有序地工作。第四代弹射座椅是作为一个技术演示项目用来展示弹射座椅技术的发展。对于第四代系统，相比于当前的座椅要求是扩展安全逃离包层的能力。要增加特殊地区的逃离包层的能力，需要从最低高度逃离，或者是以极高的速度从恶劣条件下逃离。为了实现所要求的逃离包层的扩展，方法师设计和发展了能在飞行路径中避开低海拔区域的技术，以及在不利条件下达到稳定飞行并且飞行员能在高速逃离中得到保护的技术。第四代演示系统由一系列先进的技术系统组成：可控推进、数字飞行控制和高速保护装置。

The ejection seat is an important part of emergency survival system in airplane.

When a crewmember must leave the aircraft in an emergency, it is the ejection seat that lifts him or her safely out of and away from the aircraft. In most cases, if a pilot found himself in trouble in the 20s, it was relatively easy to simply disengage the seat harness and jump over the side of the machine so that the parachute could be deployed for a safe descent. However, as the power and performance of aircraft continued to increase, it quickly became apparent that the risks involved in simply jumping over the side of an aircraft were significant. The ejection seat was important equipment to safe the crewmember life in modern aircraft in an emergency. The ejection seat system then deploys the parachute that will slowly return the crewmember to Earth. Ejecting from an aircraft takes less than two seconds from the moment the crewmember pulls the ejection handle until a parachute begins to unfurl overhead. For the ejection seat to save a crewmember's lift, nearly 500 parts in the seat's system must function properly and quickly in sequence. The Fourth Generation Ejection Seat was developed as a technology demonstrator program to showcase some of the technologies developed for the CREST Seat. For a fourth generation system, the requirement is to expand the safe escape envelope relative to that of current seats. The specific areas of the escape envelope in which increased capability is required are for escape under low altitude, adverse attitude conditions and for escape at extremely high speed. In order to achieve the required expansion of the escape envelope, the approach was to design and develop technologies which would achieve control of the flight path for ground avoidance in the low altitude, adverse attitude conditions and which would achieve stable flight and crewmember protection in escape at high speed. The fourth generation demonstration system consists of a set of advanced technology systems: controllable propulsion, digital flight control and high speed protection devices.

缩略语

ACES　Advance Concept Ejection Seat　先进概念弹射座椅

DRS　Digital Recovery Sequencer　数字化序列器

DRI　Dynamic Response Index　动态响应指数

DEP　Design Eye Position　设计眼位

HEED　Helicopter Emergency Egress Device　直升机应急逃生设备

MAXPAC　Multi-Axis Pintle Attitude Control　多轴针栓姿态控制

MDRC　Multi Axial Dynamic Response Criteria　多轴动态响应指数

NACES　Navy Aircrew Common Ejection Seat　海军飞行员通用弹射座椅

SMDC　Shielded Mild Detonating Cord　铠装柔性导爆索

STAPAC　Seat Trajectory and Pitch Attitude Control　座椅轨迹和俯仰姿态控制

TVC　Thrust Vector Control　推力矢量控制

专业术语

𝒜 a

安全舱门　safety hatch
安全措施　safety measure
安全带　safety harness
安全带/固定带　life/safety/securing belt
安全带扣环　safety-belt buckle
安全服　safety clothing
安全高度　safe altitude
安全环　safety link
安全间隙　safe clearance
安装角/迎角/入射角　angle of incidence/incidence angle
安全救生包线　safe escape envelope
安全开伞速度　safe parachute deployment/opening speed
安全设备　safety equipment
安全锁　safety lock
安全弹射包线　safe ejection envelope
安全弹射包线显示系统　safe ejection envelope display system
安全系数　safe coefficient/factor
安全压力膜盒　safety pressure aneroid
安全应急离机能力(性能)　safe escape capability
安全装置/保险机构　safety assembly

ℬ b

拔销机构　sear withdrawal unit
包线　envelope curve
包线外弹射　out-of-envelope ejection

保险销　arming pin/safety pin
保险装置　deactivating device
保障设备　support equipment
爆破试验　burst test
爆炸减压条件　explosive decompression condition
爆炸索　explosive cord
备用营救系统　standby rescue system
背带调节扣　harness strap adapter
背带解脱机构　harness release mechanism
背带解脱手柄　harness release lever
背带解脱子系统　harness release subsystem
背带快速调节扣　quick-fit harness adapter
背带连接带　harness tying tape
背带连接扣　harness adapter
背带锁　harness lock
背带系统　harness assembly
背带系统解脱机构　harness release actuator
背带自动解脱系统　automatic harness release system
背垫　back cushion/back-pad
背伞系统　piggyback system
背式伞包　back-pack parachute
背胸方向过载　back-to-chest/prone/transverse over loading
便携式生命保障系统　portable life support system
标准高度　standard altitude
表速/指示空速　indicated airspeed
并(列双)座弹射　side-by-side ejection
并列式多级火箭　multistage cluster rocket
并列双座飞机　side-by-side two-seater
并座座舱　side-by-side cockpit
不利姿态　adverse attitude

C c

舱(口)盖抛放机构　hatch jettison actuator
舱盖分离机构　canopy release
舱盖粉碎系统　canopy fragilization subsystem
舱盖锁　canopy lock
操纵(控制)手柄　control handle
侧滑角　angle of sideslip
侧向轨迹发散子系统　lateral trajectory dispersion subsystem
长伞衣套　full deployment bag
常规飞行服　regular flight cloth
敞开角　open angle
敞开式弹射座椅　open ejection seat
车载设备　sled-mounted equipment
乘员(工作)位置　crewstation
乘员/机组成员　crewmember
乘员/空勤人员　crewman
乘员生命保障设备　aircrew life support equipment
程序弹射　sequenced ejection
程序弹射系统　sequenced ejection system
程序控制(机构)　sequence control
程序系统　sequencing system
持续过载　prolonged/sustained G
充气浮囊　air-filled flotation bag
充气阶段　inflation phase
充气救生船　pneumatic raft/life raft
充气时间/张满时间　fill time
充气式救生船　inflatable liferaft
充气式坐垫　inflatable seat cushion
出舱阶段　tip-off phase
出入口/检查口　access hatch

穿(列双)座弹射　tandem ejection

穿盖　penetration of canopy

穿盖弹射　through-the-canopy/ejection through the canopy

穿盖器/穿盖装置　canopy penetrator/canopy breaker

穿盖载荷　canopy penetration load

传爆系统　igniter train

串列布局/串座布局　tandem arrangement

串列式多级火箭　multistage tandem rocket

串列双座飞机　tandem two-seater

串座(布局的)飞机　tandem seated aircraft

串座座舱　tandem cockpit

垂直/短距起落飞机　V/STOL aircraft

垂直过载　vertical G-force

垂直起落飞机　VTOL aircraft

垂直速度　vertical velocity

春秋飞行装备　intermediate flying assembly

D　d

达特稳定系统　DART stabilization system

打火操纵连杆　firing control linkage

打火手柄　firing handle

打火手柄(机构)　firing handle assembly

大过载　heavy/high G-force

代偿服　counterpressure clothing

单点快卸锁　single-point release buckle

单点联合背带　single-point combined harness

单路输出点火机构　single output initiator

单人氧气装置　individual oxygen apparatus

单伞系统　single parachute system

单座弹射　solo ejection

弹道曲线　ballistic curve

弹道式弹射座椅 ballistic-type/cartridge-fired ejection seat
弹道最高点/轨迹顶点 apex of trajectory
弹药起爆器/弹药点火器(机构) cartridge initiator
弹药作动装置 cartridge-actuated device
当量(座舱)高度 equivalent altitude
当量空速/等效空速 equivalent airspeed
导轨 guide track/rail
导火索定时器 powder-train timer
导向滑轮 guide roller
倒飞 inverted/upside down flight
倒飞状态 inverted flight condition
倒飞状态的飞机 inverted aircraft
倒飞姿态 inverted attitude/position
登船(机)扶手 boarding handle
低空 low altitude
低空弹射 low altitude ejection
低空弹射救生能力(性能) low level ejection capability
低空倒飞救生能力(性能) low altitude inverted capability
低空跳伞 low-altitude bailout
低空应急救生性能 low level escape performance
低空应急离机能力(性能) low altitude escape capability
低速 lower airspeed
低速/不利姿态(条件下的)弹射 low speed/adverse attitude ejection
低速弹射 low speed ejection
低速弹射条件(状态) low-speed ejection condition
地面保险销 ground safety pin
地面弹射 ground-level ejection
地面动态/地面有速度试验 ground dynamic test
地面高度 terrain altitude
地面静态试验 ground static test
地面拉直试验 ground deployment test

地面离机　ground egress
地面应急离机　ground-level escape
地勤人员　ground crew
第 3 百分位数的乘员　3rd percentile crewman
点火控制机构　ignition control
点火器/燃爆机构　initiator unit
点火延迟　firing/ignition (time) delay
电动调节机构(制动器)　electrically powered adjusting actuator
电起爆器　electric initiator
电子延时(机构)　electronic time delay(s)
定时系统　timing system
定位销　locating pin/locating plunger
动力减震器　dynamic vibration absorber
动态/有速度弹射试验　dynamic ejection test
动态响应指数　dynamic response index
动载　dynamic load
多级火箭　multistage rocket
多人救生船　multiseater life raft
多用途飞机　multi-purpose/multi-role aircraft
多座飞机　multi-crew/multi-place aircraft

F f

发散轨迹　divergent path
翻滚角速度　tumbling rate
反推力火箭缓冲系统　retrorocket impact attenuation system
方位角　directional angle
防暴露服/飞行抗浸服　anti-exposure flying coverall
防弹衣　armored clothing
防寒救生服　cold weather survival clothing
防护服　protective clothing
防火飞行服　fire resistant flight clothing

防火服　fire protective garment
防火装备　chemical defence gear
防生化服　CB protective clothing
防水服　waterproof garment
飞机滑跑开伞试验　aircraft taxi deployment test
飞行(中)弹射　in-flight ejection
飞行安全设备　air safety equipment
飞行弹射试验/空中弹射试验　in-flight ejection test
飞行弹射座椅应急离机方案　fly-away escape concept
飞行服　flight/flying gear
飞行服/飞行衣　flight clothing
飞行轨迹角　flight path angle
飞行开伞试验　fight deployment test
飞行口粮　flight ration
飞行员装备　aircrew equipment assembly
飞行中离机　in-flight egress
分离机构　separation mechanism
分离阶段　separation phase
分离式伞衣套(开伞袋)　free deployment bag
分离速度　separation velocity
分离子系统　separation subsystem
分时机构/定时机构　timing device
风洞　wind tunnel
风洞试验　blower/wind tunnel test
风向火箭发动机　WORD motor
浮囊　floatation bag
俯冲角　dive angle
俯仰/横滚姿态　pitch/roll attitude
俯仰角速度　angular velocity in pitch/pitching velocity
俯仰控制机构　pitch control
俯仰控制装置　pitch control unit

俯仰稳定系统　pitch stabilization system

俯仰稳定装置　pitch stabilization unit

俯仰稳定子系统　pitch stabilization subsystem

负过载　negative G force

G　g

高度控制机构　altitude control mechanism

高度时间释放机构　altitude time-release mechanism/uint

高度特性曲线　altitude characteristic curve

高空　high altitude

高空弹射　high altitude ejection

高空防护装备　high altitude protective assembly

高速弹射　high speed ejection

高速弹射条件(状态)　high-speed ejection condition

高速气流　high speed airstream

高速跳伞　high-speed bailout

告警灯　warning light

个人防护装备　personal protection equipment

个人救生包　personal survival pack

个人装备　personal equipment/personal gear

固定(拉)绳开伞　static-line deployment

管状药柱　tubular grain

惯性卷轮操纵手柄　inertia reel control handle

惯性锁紧机构　inertia-locking mechanism

惯性锁紧卷轮机构　inertia lock reel mechanism

轨迹包线　trajectory envelope

轨迹顶点　vertex of trajectory

轨迹发散火箭　trajectory divergence rocket

轨迹发散子系统　trajectory divergence subsystem

轨迹发散阻力器　divergence thruster

轨迹高度　trajectory altitude/height

轨迹角　trajectory angle
轨迹曲线　path curve/curve of trajectory/trajectory history
过载耐受极限　G tolerance
过载增长率　rate of acceleration onset

H h

海拔高度/海平面高度　sea-level altitude
海上救生(打捞)设备　salvage equipment
海上救生背心　sailing jacket
航空工效学　aviation ergonomics
航空生物学　aerobiology
横滚角　roll angle/angle of roll
横滚角速度　angular velocity in roll/rolling velocity
横滚角速度　roll rate/rolling rate
横向过载　transverse G-load
后(座)舱　after cockpit/rear cockpit
滑出　taxi-out
滑跑(行)　taxing/taxi-run
滑梯/救生船　combination slide/life raft
环形弹性吸能垫　annular resilient shock absorber
缓冲囊　impact bag
回收系统　recovery system
回收装备　recovery gear
火工品部件　pyrotechnic component
火工品定时器　pyrotechnical timer
火工品切割器　pyrotechnic cutter
火工品切割器　explosive-actuated guillotine
火箭包　rocket motor/pack
火箭包远距点火器　remote rocket package initiator
火箭包远距点火系统　remote rocket firing system
火箭车试验滑轨　rocket sled test track

火箭弹射座椅　rocket assisted seat/rocket ejection seat
火箭动力软着陆系统　rocket powered soft landing system
火箭滑车　rocket sled/rocket propelled sled
火箭滑车试验　rocket-sled test
火箭滑轨　rocket track
火箭喷流　rocket blast/efflux
火箭牵引装置　rocket extractor
火箭助推弹射　rocket-assisted ejection
火箭助推器　rocket assist
火药作动抛放机构　explosive operated jettison mechanism
货舱　cargo compartment

J j

机电式执行机构　electronic-mechanical actuator
机械打火机构　mechanical firing unit
机组/飞行人员　flight crew
极限高度 extreme altitude
急救　first aid
急救包　first aid kit
急救设备　first aid apparatus/equipment
集体跳伞　mass jump
加速度时间曲线　acceleration time curve
加压呼吸状态/加压供氧状态　pressure breathing condition
夹角　included angle
假人弹射　dummy ejection
假人弹射试验　dummy test ejection
驾驶舱　crew compartment
肩带　shoulder safety belt/shoulder harness
肩带调节手柄　shoulder-harness control lever
肩带拉紧机构　shoulder harness take-up mechanism
肩带拉紧手柄　harness retraction handle

肩带拉紧装置　shoulder harness take-up device
肩带强制拉紧机构　harness powered retraction unit
减速/时间包线　deceleration/time envelope
减速系统　retardation/retarding system
减压器适配器　pressure-reducer adapter
剪切机构/切割器组件　guillotine assembly
剪切销　shear/shearing pin
降落/着陆(水)　alight
降落伞　chute
降落伞背带(系统)　parachute harness
降落伞开伞绳挂环　parachute lanyard anchor
降落伞牵引(拉出)　parachute extraction
降落伞箱　parachute box
降落伞照明弹　parachute light
角加速度　angular acceleration
角速度　angular velocity
铰链　hinge
径向过载　radial G-force
静态/零速度弹射试验　static ejection test
静态开伞试验　static opening test
静态空投试验　static drop test
救生(的)　life-saving
救生包　survival equipment package/survival equipment container/survival kit
救生包/应急氧气系统　survival kit and emergency oxygen system
救生包启动手柄　survival kit actuator handle
救生包系统　survival kit system
救生背心　life-jacket, life vest type preserver, flotation/life jacket, life saving waistcoat
救生成功率　survival success rate
救生船　crash boat/life boat

救生筏　liferaft

救生船自动充气机构　automatic liferaft inflation unit

救生电台/人员报位信标　survival radio/personnel locator beacon

救生率　survival rate

救生能力　lifesaving capability

救生伞　escape chute

救生须知　reference-survival

救生氧气瓶　survival bottle

救生用品　survival gear

救生装备/救生物品　survival equipment

救生装置/救生设备　life saving apparatus/device

K k

开关手柄　on-off control

开口销　cotter/splint pin

开伞(拉直)　deploy

开伞(拉直/展开)高度　deployment altitude

开伞/主伞展开　parachute deployment

开伞定时装置　chute/parachute timer

开伞高度　opening altitude/altitude of deployment

开伞机构　chute development mechanism

开伞角　deployment angle

开伞拉环　ripcord handle

开伞拉绳(环)　stack deployment line/parachute rip cord

开伞速度　parachute opening speed

开伞条件　opening condition

开伞载荷　opening load

开伞周期　opening cycle

抗暴露服/抗浸服　anti-exposure assembly

抗荷服　antigravity garment/G-suit

抗浸服(海上)/救生服　survival garment
靠背　backrest
靠背切线　back tangent line
靠背释放手柄　backrest release handle
空间轨迹发散子系统　spatial separation subsystem
空降/空投　airland
空气缓冲器　air buffer
空勤人员　air crew
空速表　air-speed gauge
空投后鉴定试验　post-drop evaluation test
空投试验　airdrop test
空中弹射　airborne ejection
空中碰撞　air/midair collision
空中碰撞/空中失事/坠毁/摔机　aircrash
空中事故/飞行事故　inflight emergency
空中应急离机　airborne escape/in-flight escape
空中营救用具　air rescue kit
快速分离机构　quick-release means
快速开伞　rapid deployment
快速释放扣/快卸锁　quick-release buckle
快卸接头　quick-disconnect
快卸锁　quick release box/lock
快卸销　quick-release pin

L l

拉环　rip-cord grip
拉紧系统　tiedown system
拉直冲击载荷　deployment shock load
拦阻设备　arresting gear
离地高度　terrain clearance/terrain clearance altitude
离地事故　liftoff emergency

离机速度　exit velocity
离机系统　egress system
连接带　lanyard
连接绳拉直(法)　bridle deployment
联锁机构　interlocking device
临界高度　critical altitude
临界开伞高度　critical parachute open altitude
零/零弹射试验　zero-zero ejection test
零/零救生能力　null-null survival ability
零高度　zero altitude
零秒开伞拉绳　zero second parachute deployment lanyard
零秒延迟开伞(机构)　zero-second parachute delay
零速　zero airspeed
六自由度　six degree-of-freedom
陆上跳伞　overland bailout
旅客舱　passenger compartment
旅客机　passenger-carrying airplane
落点救援　on-scene assistance

M m

密闭舱式应急离机系统　cockpit capsule type escape system
密闭弹射座舱　encapsulated capsule
密闭式弹射座椅　encapsulated/enclosed ejection seat
密闭式救生船　encapsulating life raft
密闭式应急离机系统　encapsulated escape system
密闭座舱　closed cockpit/enclosed cockpit
面帘式(弹射)打火操纵机构　face-curtain/face screen firing control
面帘式(弹射)打火手柄　face-blind firing handle
面帘式(弹射手柄)　face curtain handle
面罩压力调节器膜盒　mask pressure regulator aneroid
模型试验　model test

膜盒机构　aneroid mechanism

膜盒气压计/真空膜盒　aneroid/capsule aneroid

P　p

爬出座舱跳伞　over-the-side bailout

爬升率　rate of climb

排放活门制动器　outflow valve actuator

旁侧式弹射手柄　side-mounted ejection control handle

抛盖火箭　canopy remover rocket

抛盖机构　hood jettison gear/canopy jettison mechanism

抛盖手柄　canopy jettison handle

抛盖栓　canopy jettison breech

抛盖子系统　cockpit canopy removal subsystem

抛盖作动筒　canopy jettison jack

抛伞　cut-away

跑道高度弹射　runway level ejection

偏航轨迹发散系统 yaw divergence system

偏航角　angle of yaw

偏航角速度　angular velocity in yaw/yawing velocity

偏航角速度　yaw/yawing rate

偏斜角　offset angle

偏心角　eccentric angle

漂浮角　flotation angle

平飞弹射　zenith hemisphere ejection

平飞状态　upright position

平飞状态的飞机　upright aircraft

平飞姿态/正常姿态　normal attitude

平均待救时间　mean-time-to-rescue

平均无故障时间　mean cycle between failures

迫降　compulsory/forced landing

Q q

起爆　explosive initiation

起飞事故　takeoff emergency

气垫减震器　air cushion shock absorber

气动力充气装置　aerodynamic inflation aid

气动载荷　aerodynamic load/air load

气密座舱　air-tight cabin

气囊　air bag

气压表　barometer gauge

气压制动器　pneumatic actuator

牵引火箭　tractor rocket

牵引伞　extract parachute

牵引式应急离机系统　extraction escape system

牵引式座椅　extraction seat

牵引系统　extraction system

前(座)舱　front cockpit

强制跳伞　force/forced jump

切割器　severance cutter/unit

切割系统　severance system

切线角　tangent angle

倾斜角　inclination angle

躯干约束系统　torso restraint system

全天候救生船　all weather life raft

全自动弹射座椅　fully automatic ejection seat

R r

燃气(压力)作动机构　gas pressure actuator

燃爆件/爆炸品/火工品　explosive component

人/椅分离器　butt-snapper/man-seat separation device

人/椅分离火箭　man/seat separation rocket
人/椅分离系统　seat/man separation system
人/椅分离子系统　man/seat separation subsystem
人体约束子系统　personal restraint subsystem
锐角　angular angle

S s

三管弹射筒　three tube ejection gun
伞包　parachute pack
伞包弹簧带/开包带　pack opening band
伞包组件　pack assembly
伞顶　apex of parachute
伞绳缠绕　suspension line twist/line twisting
伞绳拉直　suspension line deployment
伞绳拉直载荷　line stretch load
伞绳先拉直/先伞绳拉直法　lines-first deployment
伞投放试验塔　drop test tower
伞衣过早脱出　premature emergency of canopy
伞衣套/(开)伞袋　deployment bag
伞衣套拉直(法)　bag deployment
伞衣透气量　parachute porosity
伞衣脱离锁　canopy-release assembly
伞衣先拉直/先伞衣拉直法　canopy-first deployment
伞衣应急解脱机构　parachute canopy emergency release
伞衣载荷/舱盖载荷　canopy loading
沙漠救生技术　desert survival technique
上反角/二面角　anhedral angle/dihedral angle
射伞枪　parachute gun
射伞枪弹头连接带　withdrawal line
升限　ceiling altitude
生存待援阶段　survival phase

生存方法　survival technique
生存机会　survival potential
生存口粮　survival ration
生存物品/救生物品　survival aid
生存物品包　hit-and-run kit
生存训练　survival training
生活舱　maintenance compartment
生理安全高度　physiological safe altitude
生理安全状态　physiologically safe condition
生理当量高度　physiologically equivalent altitude
生理鉴定　physiological assessment
生理需要　physiological demand
生命保障　life-support
生命保障系统　life-support system
生命保障子系统　life support subsystem
失事高度/毁坏高度　abort altitude
时间释放机构　time-release mechanism/unit
使用寿命　active life
视角　angle of view
试飞/飞行鉴定　flight evaluation
适航的　airworthy
适航性要求　airworthiness requirements
适摔性/耐坠毁性　crash-worthiness
手柄　hand grip
手控跳伞　manual bailout
摔机着陆碰撞　crash landing impact
双基药柱　double-base grain
双路输出点火机构　dual output initiator
双座程序弹射　dual sequence ejection
双座串列布局　dual-tandem arrangement
双座弹射　dual ejection

双座的　two-place/two-seat

双座飞机　two-seater

双座座舱　two-crew cockpit

水平飞行　horizontal flight/level flight

水平姿态　level attitude

水上救生服　floatation garment

水水迫降　ditching

水上迫降撞击　ditching impact

水上跳伞　over water bailout

水下弹射　underwater ejection

水下呼吸装置　underwater breathing apparatus

水下应急离机　underwater escape

水下应急离机能力试验　underwater escape capability test

水下应急离机子系统　underwater escape subsystem

四肢约束系统　limb restraint/retention system

搜索与营救信标设备　search and rescue beacon equipment

锁弹机　breech firing unit/primary breech

锁定机构　locking gear

T t

弹射(操纵)手柄　ejection control handle

弹射包线　ejection envelope

弹射操纵机构　ejection control mechanism

弹射操纵手柄　ejection control handgrip

弹射成功率　ejection success rate

弹射程序控制子系统　ejection sequencing subsystem

弹射导轨　ejection/ejection guide rail

弹射导轨(安装)角　ejection rail angle/rail angle

弹射高度　ejection altitude

弹射轨迹　ejection path/ejection trajectory

弹射机构　ejection mechanism

弹射间隙　ejection clearance
弹射阶段　ejection phase
弹射救生成功率　ejection survival rate
弹射起动　ejection initiation
弹射气动载荷　ejection airload
弹射器/弹射　catapult
弹射器启动装置　catapult trigger
弹射前的姿势　pre-ejection position/posture
弹射时的姿势　ejection position/posture
弹射试验　ejection test
弹射试验塔　ejection test tower
弹射手柄　ejection grip/ejection handle
弹射速度　ejection velocity
弹射跳伞　ejection bailout
弹射筒　ejection gun
弹射筒打火机构　ejection gun firing unit
弹射筒内筒　inner piston
弹射筒中筒　intermediate piston
弹射载荷　ejection load
弹射指令　ejection command
弹射装置　ejection equipment
弹射准备动作/弹射前准备　pre-ejection action
弹射准备阶段　pre-ejection phase
弹射座舱　ejectable cockpit/crew module
弹射座椅　ejection seat
弹射座椅型应急离机系统　ejection seat type escape system
套筒式弹射筒　telescopic ejection gun
套筒式弹射装置　telescope-tube ejector
调温服　thermoconditional clothing
跳伞　bailing-out/bailout/bail-out
跳伞(安全)门　bailout/jump/parachute door

跳伞舱口　parachute exit
跳伞舱门　bailout hatch
跳伞舱门开伞　jump-door deployment
跳伞程序　bail-out procedure
跳伞服　jumping garment
跳伞高度　bail-out/jumping altitude
跳伞供氧调节器　bail-out oxygen supply regulator
跳伞供氧器　bail-out oxygen
跳伞信号灯　bailout light/jumping light
跳伞训练　jump/parachute training
跳伞氧气瓶　bailout bottle/cylinder
跳伞装备软管　parachute unit hose
通风服　ventilating/ventilation garment,air ventilated clothing
通风外套　air ventilated coverall
头部损伤标准　head injury criteria
头脚方向过载　head-toe G
头靠(安装)角　headrest angle
头靠垫/头垫　headrest cushion/head cushion
头靠调节手柄　headrest adjustment lever
头靠伞箱　headrest box
投放(空投)高度　drop altitude
推力角　thrust angle
推力矢量控制(机构)　thrust vector control
陀螺微调火箭　gyro-controlled vernier rocket
陀螺微调火箭俯仰稳定子系统　gyrocontrolled vernier rocket pitch stabilization subsystem
陀螺旋转执行机构　gyro spin-up actuator

U　u

U 形销　clevis pin

W w

外观检查　appearance test
微型爆破索　miniature detonating cord
微调火箭发动机　vernier rocket motor
稳定伞连接带　drogue bridle
稳定(减速)伞　drogue chute
稳定伞　stabilizing chute
稳定伞伞包/稳定伞伞箱　drogue container
稳定装置　stability/stabilizing device
稳定伞射伞枪　drogue gun
稳定伞双弹射伞枪　dual cartridge drogue gun
稳定伞的开伞载荷　drogue load
稳定减速伞　drogue parachute
稳定伞射伞枪弹头　drogue-gun piston
稳定子系统　stabilization subsystem
稳定性试验　stability trial
稳定/减速子系统　retardation and stabilization subsystem
稳降　descending at steady rate/stabilized descent
稳降阶段　steady-state descent phase
稳降速度　steady-state descent rate
稳态下降　steady-state descent/steady fall

X x

吸能器　impact shock absorber
吸能装置　energy absorber
系留式伞衣套(开伞袋)　controlled deployment bag
下沉率　sink rate
下沉速度　sinking velocity
下滑角　glide angle
限臂系统　arm restraint system

限腿带　leg belt

限腿带的箍带　leg restraint garter

限腿机构　leg restraint gear

限腿系统　leg restraint system

限位带　restraint harness

相对气流　relative airstream

向上弹射　upright ejection

向下弹射　down ejection/downward ejection

向下弹射座椅　downward ejection seat

信号传输子系统　signal transmission subsystem

信号装置　signaling aid

性能包线　performance envelope

性能曲线　performance curve

性能指数　performance index

胸背方向过载　chest-to-back G/supine/transverse A-P G

旋翼机　rotary-wing aircraft

旋转角　angle of turn

旋转速度　rotational rate

巡航高度　cruising altitude

巡航空速　cruising airspeed

Y　y

延迟点火　delayed ignition

延迟机构　delay unit

延迟开伞　parachute delay/delayed parachute deployment

延时部件　delay component

延时点火器　delay initiator

延时动作(作用)　time-lag action

延时机构　delay time mechanism/time-delayed mechanism

延时膜盒　time delay bellows

延时引爆器　time delay initiator

延时装置 delay device
样机全尺寸试验 mock-up test
腰带 waist belt
夜间跳伞 night jump
液冷服 liquid cooling garment
液压作动机构/液压作动筒 hydraulic actuator
腋下救生器 under-arm life-preserver
移动角 angle of excursion
椅背(安装)角 seat back angle
椅背火箭 seat back rocket
椅盆 seat bucket
椅盆调节手柄 seat-pan adjusting lever
椅盆升降机构(制动器) seat pan actuator
椅盆升降手柄 seat pan raising handle/seat raising lever
椅盆组件 seat pan assembly
椅装点火器 seat-mounted initiator
椅装设备 seat-mounted equipment
椅装应急氧气系统 seat mounted emergency oxygen system
引导伞 pilot chute/extraction parachute/extractor parachute
应急舱门 egress/emergency/escape hatch
应急舱门系统 exit system
应急出口 emergency exit
应急穿盖系统 through canopy system
应急措施 emergency measure
应急措施(方法) emergency approach
应急弹射 emergency ejection
应急弹射离机 ejection escape
应急动作训练 emergency drill
应急高度 emergency altitude
应急解脱(分离/释放)手柄 emergency release handle
应急解脱操纵机构 emergency release control

应急解脱机构　emergency release
应急救生包　bail-out/egress kit
应急救生包线　escape envelope
应急救生舱　emergency escape capsule
应急救生伞　emergency/emergency escape parachute
应急口粮　emergency ration
应急拉环　emergency rip-cord grip
应急离机　emergency egress
应急离机出口　escape exit
应急离机的临界高度　critical escape altitude
应急离机高度　escape altitude
应急离机告警系统　escape alert system
应急离机轨迹　escape trajectory
应急离机滑梯　escape/jump slide
应急离机能力　escape capability
应急离机事故　escape emergency
应急离机通道　escape path
应急离机通道清除子系统　escape path clearance subsystem
应急离机系统　emergency escape system
应急离机系统操纵机构　escape system control
应急离机系统的设计校核试验　escape system design verification test
应急离机系统的座舱安装/协调试验　escape system cockpit fit and compatibility test
应急离机系统结构(布局)　escape system configuration
应急离机系统试验　escape system test
应急离机训练　escape drill
应急离机装备　emergency escape equipment
应急抛盖系统　canopy jettison system
应急漂浮装备　emergency floatation gear
应急情况分析　emergency analysis
应急设备　emergency apparatus/equipment

应急使用　emergency application

应急手柄　emergency manual handle

应急条件/应急情况　emergency condition

应急着陆　emergency landing

(应急)分离舱/弹射座舱　capsular/capsule/capsule-type cabin

营救设备　rescue/aid equipment

有效透气量　effective porosity

约束系统　restraint system

阻力减速伞　drag parachute

Z z

折叠式救生船　collapsible life boat

真空试伞塔　exhaustible parachute tower

真人弹射　live ejection

真人弹射塔试验　live subject ejection tower test

真人试跳　live jump test

真人跳伞　live drop

真速　actual airspeed/true airspeed

正常开伞　routine deployment

正过载　positive G force

止动销　locking pin

指令弹射　command ejection

指令弹射系统　command ejection system

指令弹射装置/指令弹射机构　command ejection facility

指令选择控制机构　command selection control

指令选择控制子系统　command selection control subsystem

中高速试验　medium/high speed test

中空　medium altitude

中央弹射/操纵手柄　center ejection control handle

钟表延时　clock work time delay
重心范围　center-of-gravity envelope
重心位置　condition center-of-gravity location
弹射操纵机构　primary ejection control
主点火器　primary igniter
主回收伞子系统　main recovery parachute subsystem
主火箭发动机　main rocket
主伞　main chute
主伞伞包　main canopy pack
专用设备　special equipment
撞击试验　bump test
撞针　firing pin
撞针行程　firing pin travel
准备出机　poised exit
着陆角　landing angle
着陆事故　landing emergency
姿态参照系统　attitude reference system
姿态控制(系统)　attitude control
自动充气装置　automatic inflation device
自动开伞　automatic deployment
自动开伞器　automatic chute opening device
自动开锁器　automatic unlocking device
自动应急离机系统　automatic escape system
自给式呼吸装置　self contained breathing apparatus
自给式水下呼吸装置　self contained underwater breathing apparatus
自行开伞/意外开伞　spontaneous deployment
自由拉直　free deployment
自由跳伞离机高度　egress altitude for free jumps
自由坠落阶段　free descent phase

总装组件　top assembly
阻力伞　drag chute
最大安全跳伞空速　jump airspeed
最大飞行高度/动升限　peak altitude
最大过载　maximum G-force
最大开伞动载　peak opening shock
最大拉直力　peak snatch force shock
最低安全弹射高度　minimum safe ejection altitude
最低安全高度　minimum safe altitude
最低安全空投高度　minimum safe dropping altitude
最低高度　minimum altitude
最佳轨迹　optimum trajectory
最佳姿态　optimum attitude
作动机构/传动装置　actuation gear
座舱(压力)高度　cabin/cockpit pressure altitude
座舱爆炸减压　explosive loss of cabin pressure/blowout
座舱盖　cockpit canopy
座高调节机构　height/seat adjustment actuator
座高调节手柄　vertical seat adjustment lever
座式(伞/救生)包　seat pack
座式救生包　seat type survival kit
座椅安装导轨　seat attachment rail
座椅参考点　seat reference point
座椅弹射导轨　seat-ejection rail
座椅弹射手柄　seat ejection lever
座椅弹射训练　seat-ejection training
座椅导轨　seat rail/seat track
座椅滑轮　seat roller
座椅加速阶段　seat boost phase

座椅联动程序系统　interstate sequencing system

座椅行程　seat travel

座椅中央打火手柄　center seat firing handle

座椅组件　seat assembly

(座椅弹射后的)越尾距离　tail clearance

供氧系统

导　语

供氧系统应该适用于飞机的任务要求。供应类型包括气态氧、液态氧和机载氧气生成系统。氧气系统应与飞行器的操作环境兼容并且可操作。系统组件也应符合环境存储要求。飞机乘员的数量和类型是一个基本的功能要求,并且能在初始飞机氧气系统中被提到。功能子系统由个人类型提供,并且必须使用氧气子系统。与这些类型的人员相关联的性能特征很大程度上依赖飞机及其任务。氧气系统应该在飞机预定的正常或者紧急任务中供给所有乘员以及其他人。氧气系统由以下功能子系统组成:

① 乘员呼吸系统;

② 伞兵氧气系统;

③ 任务专家氧气系统;

④ 航空医用氧气系统;

⑤ HALO/HAHO 氧气子系统;

⑥ 旅客氧气系统;

⑦ 紧急乘员氧气子系统;

⑧ 手动跳伞氧气;

⑨ 氧气相关组件;

⑩ 飞机消防员便携式装配;

⑪ 直升机紧急逃生装置;

⑫飞机压力服供给。

每个人有责任与民用、军用、商用以及通用航空的氧气设备联系起来,至少了解关于呼吸的基本事实是先决条件。呼吸是一个生物获得其必需的细胞功能(代谢)所必要的氧气的过程,以及排出执行功能的产出(主要为二氧化碳)的过程。在人体的这个过程中包括通风(吸入/呼出),氧气从肺部到血液的扩散(以及二氧化碳从血液到肺部),

肺部和组织间的血液循环，氧气从血液到组织的扩散（以及二氧化碳从组织到血液）。

氧气系统的设计目标是尽量减少正常呼吸的障碍。呼吸的工作量只占身体总能量消耗的一小部分。由氧气设备对呼吸的自动生理机制施加的任何额外负荷都会干扰正常的呼吸模式。

The oxygen system supply shall be appropriate for the mission requirements of the aircraft. Supply types include gaseous oxygen (GOX), liquid oxygen (LOX), and on-board oxygen generating systems (OBOGS). The oxygen system shall operate and be compatible with the operational environment of the air vehicle. System components shall also meet environmental storage requirements. The number and types of aircraft occupants is a basic functional requirement that should be called out in the initial portion of the aircraft oxygen system description. The functional subsystems are given by personal types that must use the oxygen subsystems. The performance features associated with each of these types of personnel are strongly dependent on the aircraft and its mission. The oxygen system shall support all crew members and other personnel for the normal and emergency intended missions of the aircraft. The oxygen system consists of the following functional subsystems as applicable:

① Crew breathing system;

② Paratroop oxygen system;

③ Mission specialist oxygen system;

④ Aero medical oxygen system;

⑤ HALO/HAHO oxygen subsystem;

⑥ Passenger oxygen subsystem;

⑦ Emergency crew member oxygen subsystem;

⑧ Manual bailout oxygen;

⑨ Walk-around oxygen assemblies;

⑩ Aircraft fire fighter portable assembly;

⑪ Helicopter emergency egress device(HEED);

⑫ Aircraft pressure suit provisions.

An understanding of at least the elementary facts about respiration is a prerequisite for everyone having responsibilities in connection with oxygen equipment for civilian, military, commercial, or general aviation aircraft. Respiration is the process by which a living organism acquires the oxygen necessary for its essential cellular functions (metabolism) and discharges gaseous products of those functions (primarily CO_2). In man this process consists of ventilation (inhalation/exhalation), diffusion of O_2 from lungs to blood (and CO_2 from blood to lungs), circulation of blood between lungs and tissues, and diffusion of O_2 from blood to tissues (and CO_2 from tissues to blood).

The design goal for oxygen systems is to minimize impairment of normal breathing. The work of breathing represents only a small fraction of the total energy expenditure of the body. Any additional load imposed upon the automatic physiological mechanics of breathing by the oxygen equipment will disturb normal breathing patterns. It may also cause discomfort and fatigue of the muscles involved in breathing.

缩略语

ADS　Altitude Decompression Sickness　高空减压病

AGE　Aircraft Ground Equipment　飞机地面设备

AIT　Autoignition Temperature　自燃温度

AMM　Aircraft Maintenance Manual　飞机维修手册

ANSI　American National Standards Institute　美国国家标准协会

APL　Aviation Physiology Lab　航空生理实验室

APS　Altitude Pressure Switch　高度开关

ASME　American Society of Mechanical Engineers　美国机械工程师协会

ASTM　American Society for Testing and Materials　美国材料和测试学会

ASSY　Assembly　组件

ATPD　Ambient Temperature and Pressure Dry　环境温度压力干燥状态

CBVR　Chemical Biological Visor/Respirator　化学生物战剂呼吸器

CDR　Critical Design Review　关键设计评审

CFR　Code of Federal Regulations　联邦管理规定

CGA　Compressed Gas Association　压缩气体协会

COG　Ceramics Oxygen Generator　陶瓷氧气发生器

COGS　Ceramic Oxygen Generation System　陶瓷制氧系统

CRA　Cylinder Regulator Assembly　氧气瓶调节器组件

DRR　Dispatch Reliability Rate　派遣可靠率

EEL　Emergency Exposure Limit　紧急暴露限值

EIS　Entry Into Service　进入服务

FCA　Functional Configuration Audit　功能构型审查

GOX　Gaseous Oxygen　气氧

GSE　Ground Support Equipment　地面支援设备

HALO　High Altitude Low Opening　高空跳伞-低空开伞

HALT　Highly Accelerated Life Test　加速寿命试验

HAHO　High Altitude High Opening　高空跳伞-高空开伞

HASS　Highly Accelerated Stress Screening　加速应力检验

HP　High Pressure　高压

ICD　Interface Control Drawing　接口控制图

ITM　Ionic Transport Membrane　离子传导膜

LCC　Life Cycle Cost　寿命周期费用

LOX　Liquid Oxygen　液氧

LP　Low Pressure　低压

MAWP　Maximum Allowable Working Pressure　最大允许工作压力

MSL　Mean Sea Level　平均海平面

MSOC　Molecular Sieve Oxygen Concentrator　分子筛氧气浓缩器

N/A　Not Applicable　不适用

NTPD　Normal Temperature Pressure Dry　正常温度压力干燥状态(21.1℃,101.3kPa)

NBC　Nuclear Biological and Chemical　核生化

OBONGS　On-Board Oxygen Nitrogen Generation System　机上制氧制氮系统

OBOGS　On-Board Oxygen Generating System　机上制氧系统

ODI　Overboard Discharged Indicator　机外释压指示器

PAX　Passenger　旅客

PBE　Protective Breathing Equipment　防护性呼吸设备

PCA　Physical Configuration Audit　物理构型审查

PEC　Personal Equipment Connector　个人装备断接器

PSA　Pressure Swing Adsorption　变压吸附

P/N　Part Number　件号

PODU　Passenger Oxygen Dispensing Unit　旅客氧气分配单元

PSU　Passenger Service Unit　旅客服务单元

P&T　Pressure and Temperature　压力和温度

PTT　Press-To-Test　按压测试

SDS　Space Decompression Sickness　太空减压病

SFR　System Functional Review　系统功能评估

SSR　System Requirements Review　系统需求评估

TRS　Technical Requirements Specification　技术要求规范

TSO　Technical Standard Order　技术标准条例

VGE　Venous Gas Emboli　静脉气泡

专业术语

A a

安全暴露时间　critical time of exposure

安全高度　safety altitude

安全压力　safety pressure

安全余压供氧装备　safety pressure oxygen equipment

暗视觉　scotopic vision

暗视力　darkness visual acuity

暗适应　adaptation dark

B b

板装式氧气调节器　aircraft panel mounted oxygen regulator

饱和　saturation

饱和度　saturation degree

饱和系数　saturation coefficient

保护头盔　protective helmet

爆发性高空缺氧　fulminating altitude hypoxia

暴露时间　exposure time

爆炸减压　explosive decompression

备用时间　reserve time

备用调节器　stand-by regulator

备用系统　backup system

备用氧　backup oxygen

备用氧源　secondary oxygen source

闭路供氧系统　closed circuit oxygen system

闭式回路氧气呼吸器　closed circuit oxygen breathing apparatus

闭式加压头盔　closed type pressure helmet

便携式个人肺式稀释供氧装置　portable individual demand oxygen unit

便携式供氧装置　portable oxygen apparatus

便携式氧气系统　portable oxygen system

标准供氧装备　standard oxygen gear

补充供氧　supplemental oxygen

不供氧　oxygen deprivation

不随意过度换气　involuntary hyperventilation

部分加压服　partial pressure suit

C c

侧管式部分加压服　capstan partial pressure suit

侧管式代偿背心/服　capstan pressure vest/suit

侧管式高空代偿服　capstan high altitude compensating suit

产品气　product gas

潮气量　tidal volume

乘员供氧面罩 aircrew oxygen mask

充氧活门　filling valve

充氧接嘴　filling connection

出汗率　sweating rate

纯氧　pure oxygen

重复暴露　repeated exposure

储气囊　rebreathe bag

D d

代偿/加压手套　pressure gloves

代偿　compensation

代偿背心　pressure vest
代偿区　compensatory zone
代偿压力　counter pressure
代谢率　metabolic rate
单兵供氧装备　individual oxygen equipment
单向活门　check valve
低温/冷习服　acclimatization to cold
低压舱　altitude chamber
低压差　low pressure differential
低压-减压病　hypobaric decompression sickness
低氧混合气体　low oxygen gas mixture
地面低压呼吸训练　ground pressure breathing training
地面加压呼吸训练器　ground pressure respiratory trainer
电子供氧系统　electronic oxygen system
断续式供氧调节器　oxygen demand regulator
对抗措施　countermeasure
对抗压　counter pressure
对流型通风服　convective air-ventilated suit

E e

二氧化氮　nitrogen dioxide
二氧化碳　carbon dioxide
二氧化碳产生量　carbon dioxide production
二氧化碳排出量　carbon dioxide output

F f

防护头盔　protective helmet
防护装备　protective device
防烟面罩　smoke proof mask

飞行安全　flying safety

飞行事故　aviation accident

飞行员个体防护装备　pilot personal protective equipment

飞行员工作负荷　pilot workload

非加压供氧装备　nonpressure oxygen equipment

非气密供氧面罩　unclosed type oxygen mask

肺换气量　pulmonary ventilation

肺内压　intrapulmonic pressure

肺内余压　pulmonary excessive pressure

肺泡气　alveolar gas

肺泡气二氧化碳分压　alveolar gas carbon dioxide pressure

肺泡气方程　alveolar gas equation

肺泡气气体交换　alveolar gas exchange

肺泡气氧分压　alveolar oxygen pressure

肺泡无效腔　alveolus dead space

肺式供氧　demand oxygen supply

肺式供氧装备　demand oxygen equipment

肺式加压供氧调节器　pressure-demand oxygen regulator

肺式氧气调节器　demand oxygen regulator

肺通气量　pulmonary ventilation

肺型氧中毒　pulmonary oxygen toxicity

分压　partial pressure

分子筛　molecular sieve

分子筛床　molecular sieve bed

粉尘　dust

复合式供氧装备　combined continuous oxygen equipment

G　g

干球温度　dry bulb temperature

高空代偿服　high altitude compensating suit
高空代偿服张紧装置　tentioning system of partial pressure suit
高空等效生理高度　physiologically equivalent altitudes
高空等效生理效应　altitude physiologically equivalent effects
高空供氧系统　oxygen supply system at high altitude
高空缺氧　altitude hypoxia
高空空投　altitude air drop
高空停留时间　altitude time
高空胃胀气　barometerism
高温/热习服　acclimatization to heat
高压-减压病　hyperbaric decompression sickness
个体防护装备　personal equipment
供氧等效生理高度　physiologically equivalent altitude of oxygen supply
供氧个体防护装备　personal protective equipment in oxygen supply
供氧抗荷综合调节器　oxygen supply and anti-G integrated regulator
供氧面具　supply oxygen veil
供氧面罩　supply oxygen mask/oxygen mask
供氧能力　oxygen delivery capacity
供氧等值生理高度　physiologically equivalent altitudes of supply oxygen
供氧调节器　oxygen regulator
供氧系统　oxygen supply system
供氧系统弹射试验　ejection test of oxygen supply system
供氧装备　oxygen supply apparatus
管路附件　line attachment
过饱和　super saturation
过度换(通)气　hyperventilation
过度通气晕厥　hyperventilation syncope
过载　G-force

H h

含氧百分比　percentage of oxygen

航空工程学　aeronautical engineering

航空工效学　aviation ergonomics

航空供氧防护装备应用生理学　applied physiology of aviation oxygen protective equipment

航空航天生理学　aerospace physiology

航空生理学　aviation physiology

航空生理实验室　aviation physiology lab

航空生理重点实验室　key lab on aviation physiology

航空医学　aviation medicine

耗氧率　oxygen consumption rate

黑球温度　globe temperature

呼气　exhalation/expiration

呼气活门　exhalation/expiratory valve

呼吸　breathing/respiration

呼吸频率　respiratory rate

呼吸气阻力　respiratory impedance

呼吸全脸型面罩　breathing full face mask

呼吸无效腔　respiratory dead space

呼吸性碱中毒　respiratory alkalosis

呼吸压力波动　breathing pressure fluctuation

化学供氧器　chemical oxygen supply apparatus

化学氧源　chemical oxygen source

J j

机上供氧装备　on-board oxygen equipment

机载分子筛产氧系统　molecular sieve oxygen generating system on board

急性低体温　acute hypothermia

急性高空缺氧　acute altitude hypoxia

集体供氧系统　collective oxygen supply system

集体用氧气调节器　oxygen regulator for collective use

加压背心　pressure jerkin/vest

加压供氧　pressure oxygen supply

加压供氧面罩　pressure oxygen mask

加压供氧系统　positive pressure oxygen system

加压供氧氧气调节器　positive pressure oxygen regulator

加压供氧装备　pressure oxygen equipment

加压呼吸　pressure breathing

加压呼吸性晕厥　syncope in pressure breathing

加压面罩　pressure mask

加压式呼气活门　pressure exhalation valve

加压头盔　pressure helmet

加压头盔拉紧机构　tiedown system of pressure helmet

加压治疗　compression therapy

加压座舱　pressurized cabin

假肺　simulative lung

间断性　discontinuity

间接式氧气调节器　servo-operated oxygen regulator

减压　decompression

减压时间　time of decompression

降落伞供氧调节器　survival kit oxygen regulator

救援时间　rescue time

军用运输机　military cargo-transport plane

K　k

开式供氧面罩　open supply oxygen mask

开式呼吸系统　open respiratory system

开式加压头盔　open pressure helmet

抗暴露服　anti-exposure suit

抗荷加压供氧　anti-G positive pressure oxygen supply

抗荷加压呼吸　anti-G pressure breathing

抗荷设备　anti-G equipment

抗荷设备离心机试验　centrifuge test anti-G equipment

抗荷调压器　anti-G valve

抗荷效果　anti-G effect

抗浸服　immersion suit

克裸　clo

空间等效高度　space equivalent altitude

空间实验室　space lab

空勤人员　air crew member/aircrew

空勤人员选拔　aircrew selection

空中突然失能　sudden inflight incapacitation

口鼻型供氧面罩　oral nasal type oxygen mask

跨胸廓压　trans thoracic pressure

快戴式面罩　quick donning mask

快速分离器　quick operating oxygen separator

快速佩戴式面罩　quick don mask

L l

冷水浸泡　cold water immersion

离心机　centrifuge

连接管道　connect pipe

连续供氧　continuous oxygen supply

连续流量供氧系统　continuous flow oxygen system

连续供氧装备　constant flow oxygen equipment

连续式供氧调节器　oxygen continuous regulator

连续式氧气调节器　continuous flow oxygen regulator

联合性环境负荷　combined environmental stress

露点　dew point

旅客供氧面罩　passenger oxygen mask

M m

慢性高空缺氧　chronic altitude hypoxia

弥散　diffusion

密闭服　obturation suit

密闭式供氧面罩　obturation oxygen mask/closed type oxygen mask/sealed mask

密闭头盔　closed type supply oxygen helmet

密闭座舱　sealed cabin

面罩/面具　mask

面罩挂带系统　mask harness system

面罩式氧气调节器　mask mounted oxygen regulator

模拟呼吸器　breathing simulator

膜制氧氧源　membrane generating oxygen source

N n

耐受时间　tolerance time

耐受终点　tolerance end point

囊式部分加压服　bladder partial pressure suit

囊式高空代偿服　cystiform high altitude compensating suit

脑电图　electroencephalogram

内圈大气　inner atmosphere

逆向式氧气减压器　converse flow reducer

P p

排氮　denitrogenation/nitrogen elimination/nitrogen washout

贫血性缺氧　anaemic hypoxia

平均辐射温度　mean radiant temperature

平均皮肤温度　mean skin temperature

平均体温　mean body temperature

屏气时间　breath hold time

波义耳定律加压服　Boyle's law pressure suit

Q q

气哽　chokes

气管气　tracheal air

气管气氧分压　tracheal oxygen pressure

气核　gas nucleus nuclei

气流　air movement/air flow

气流吹袭伤　blast injury

气密性　air tight sealing

气栓　gas emboli

气体的真实浓度　true concentration of gas

气温　air temperature

气压高度　pressure altitude

气氧系统　gas oxygen system

气氧源　gas oxygen source

潜水病　caisson disease

强击机　attack plane

全加压服　full pressure suit

全加压头盔　full pressure helmet

全脸型供氧面罩　supply oxygen full face mask

全压服　full pressure suit

缺二氧化碳/低二氧化碳症　hypocapnia

缺氧　hypoxia/anoxia

缺氧冲击　hypoxia pulse

缺氧后反应　post hypoxia reaction

缺氧耐力　hypoxia tolerance

缺氧耐力地面检查训练器　ground trainer for oxygen starvation endurance

缺氧试验　hypoxia test

缺氧性缺氧　hypoxic hypoxia

缺氧晕厥　hypoxic syncope

R　r

热储存　heat storage

热传递　heat transfer

热负荷　heat load

热平衡　thermal ballance

热蓄积　heat storage

热债　heat debt

人机环境　man aircraft environment

人机环境工程研究所　institute of the man aircraft environment engineering

人机环境系统　man aircraft environment system/man machine environment system

人机界面　man aircraft/machine interface

人机系统　man aircraft/machine system

溶解气体　dissolved gas

S s

瘙痒感　pruritus

身体热容量　body heat capacity

生理等效高度　physiologically equivalent altitude

生理等效效应　physiological equivalent effects

生理评定　physiological assess

生理实验　physiological experiment

生理性热暴露限值　physiological heat exposure line

生理性调节　physiological regulation

生理学要求　physiological requirement

生理训练　physiological training

生命保障模块　life support module

生命保障系统　life support system

生物转化　biotransformation

湿型抗暴露服　wet anti-exposure suit

视力模糊　vision dim/visual blurring

视敏度　visual acuity

视野　visual field

适应　adaptation

书写试验　handwriting test

舒适区　comfort zone

输氧量　oxygen delivery

双压制飞行服　two pressure flying suit

水冷服/液冷服　water cooled suit

顺向式氧气减压器　positive flow reducer

死腔　dead space

伺服式供氧调节器　serve operated oxygen regulator

T t

陶瓷制氧氧源　ceramic generating oxygen source

体表/外壳温度　shell temperature

体温过低　hypothermia

体液沸腾　ebullism

天空试验室　skylab

调节器　regulator

调压腔　pressure control cavity

跳伞供氧分系统　bailout oxygen supply subsystem

跳伞供氧器　bailout oxygen supply apparatus

跳伞供氧装备　bailout oxygen equipment/parachuting oxygen equipment

通风服　ventilation/air ventilated suit

通风增压座舱　air ventilating pressured cabin

头盔隔噪声试验　test for noise proof performance of helmet

头盔抗穿透试验　penetration resistance test of helmet

头盔碰撞试验　helmet impact test

脱饱和　desaturation

脱氧积分指数　deoxygenation integral index

W w

危险区　critical zone

温度负荷　temperature load

温度感受器　thermo receptors

无感蒸发　insensible perspiration

物理性能试验　physical performance test

X x

吸气　inspiration

吸气活门 inhalation/inspiratory valve

吸气阻力 inspiratory resistance

吸入气 inhale gas

习服 acclimatization

小余压式呼气活门 small pressure exhalation valve

心电图 electro cardiogram

胸膜腔内压 intrathoracic pressure

胸配式氧气调节器 on chest oxygen regulator

眩晕 vertigo

血/气分配系数 blood/air partition coefficient

血管迷走性晕厥 vasovagal syncope

血氧饱和度 blood oxygen saturation

血液二氧化碳解离曲线 blood carbon dioxide dissociation curve

循环停滞性缺氧 circulatory stagnant hypoxia

循环周期 cycle time

迅速减压 rapid decompression

迅速减压舱 rapid decompression chamber

迅速减压峰值 rapid decompression peak value

迅速减压试验 rapid decompression test

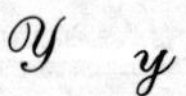

压差 pressure different

压力比 pressure proportion

压力比调节器 oxygen pressure ration regulator

氧/解离曲线 oxygen dissociation curve

氧的反常效应 oxygen paradox

氧分压 oxygen partial pressure

氧含量 oxygen content

氧耗量 oxygen consumption

氧化　oxidation

氧气操纵器　oxygen system controller

氧气传感器　oxygen sensor

氧气断接器　oxygen disconnector

氧气减压器　oxygen reducer

氧气开关　oxygen valve

氧气面罩　oxygen supply mask

氧气面罩箱　mask drop down set

氧气浓缩器　oxygen concentrator

氧气瓶　oxygen cylinder

氧气设备试验器　tester for oxygen equipment

氧气示流感应器　oxygen flow inductor

氧气示流器　oxygen flow indicator

氧气调节器　oxygen regulator

氧气系统试飞试验　oxygen system flight test

氧气卸压连接器　oxygen pressure relief connector

氧气压力表　oxygen manometer/oxygen pressure gauge

氧气余压调节器　oxygen overpressure regulator

氧气余压指示器　oxygen overpressure indicator

氧气指示器　oxygen indicator

氧容量　oxygen capacity

氧源　oxygen source

氧源系统　oxygen source system

氧源转换器　oxygen supply convertor/oxygen convertor

氧中毒　oxygen intoxication

氧烛　oxygen candle

夜间飞行　night flight

液冷服　liquid cooled garment

液态氧　liquid oxygen

液氧储氧量表　liquid oxygen quantity indicator

液氧氧气系统　liquid oxygen system

液氧源　liquid oxygen source

液氧转换器　liquid oxygen converter

一氧化碳　carbon monoxide

蚁走感　formication

椅装式氧气调节器　seat mounted oxygen regulator

意识丧失　unconsciousness

意识时间　time of consciousness

应激　stress

应急　emergency

应急加压　emergency recompression

应急氧气　emergency oxygen

用氧插座　supply oxygen socket

用氧高度　oxygen supply altitude

有效意识时间　time of useful consciousness

余压　excessive pressure

预先呼吸氧气　preoxygenation

原发性缺二氧化碳　primary hypocapnia

运动协调　motor coordination

Z z

再氮化　renitrogenation

载人航天飞行　manned spaceflight

增压/再加压　recompression

战斗机热负荷指数　fighter index of thermal stress

张紧补偿器　tightness compensator

张力　tension

障碍区　disturbance zone

蒸发型通风服　evaporative air ventilated suit

蒸气压　vapour pressure

正过载　positive G

正压呼吸　positive pressure breathing

知觉　perception

脂肪栓　fat embolism

直肠温度　rectal temperature

直接式氧气调节器　straight oxygen regulator

装备无效腔　equipment dead space

自主件节律　self rhythm

总压　total pressure

总压制度　total pressure schedule

组织中毒性缺氧　histotoxic hypoxia

最大耗氧量　maximal oxygen consumption

水/废水系统

导 语

水/废水系统供应饮用水到盥洗室和厨房，排出洗手池和马桶的废水。水/废水系统也排出门槛区域的雨水。水/废水系统包括如下子系统：饮用水、供气和污水处理。

饮用水系统为盥洗室和厨房供水。饮用水系统有如下子系统：旅客饮用水、水加热、水量指示。旅客饮用水子系统为盥洗室和厨房供水。水加热子系统加热供应到盥洗室水龙头的饮用水。水量指示子系统测量和显示饮用水子系统的水量。

污水处理系统排放来自盥洗室和厨房洗手池的水、人员如厕产生的马桶污水和舱门密封条处的雨水。污水处理系统包括如下子系统：洗涤污水、真空污水和污水箱液位指示。

洗涤污水系统有如下功能：排放来自盥洗室和厨房洗手池的水和其他液体；对盥洗室和厨房进行通风；从登机门/服务门门槛区域排放雨水。洗涤污水有如下两个子系统：污水、门槛排水。污水子系统收集来自盥洗室和厨房洗手池的洗涤废水，并通过机外排水杆排出机外。污水子系统也对盥洗室和厨房进行通风。门槛排水子系统收集来自登机门和服务门门槛区域的雨水，并将雨水通过排水接头排出机外。

真空污水系统将马桶内的如厕污水排放到污水箱中，并储存。真空马桶收集人员的如厕污水。在一个冲洗循环内，将污水冲进真空污水管路。座舱压力将污水排入真空污水箱。有如下两个能源产生污水箱的低压（负压）：真空风扇、座舱内外压力差。污水被存放在污水箱中直至地面维护时进行处理。地面维护是通过污水维护面板进行。真空污水系统有这些部件：马桶组件、污水箱、水分离器、污水箱冲洗喷嘴、污水箱冲洗过滤器、污水箱冲洗接头组件、污水箱排放阀组件、污水排放球阀、排水管路消堵阀、真空单向阀、真空风扇以及真空风扇压力开关。

污水箱液位指示系统测量和显示污水箱内污水的量。污水箱液位指示系统有如下部件：污水箱点液位传感器、污水箱连续液位传感器、逻辑控制模块和污水液位指示灯。污水箱液位指示系统有如下功能：监控和显示污水箱的液位；当污水箱满时，停止马桶的工作；当污水箱满时，盥洗室停止工作，给出指示；当液位传感器脏污时，给出指示；系统自检(BIT)。

水箱增压系统对清水箱进行增压。使用来自气源系统或压气机的压力对水箱增压。水箱增压系统有如下功能：控制增压空气进入清水箱，选择增压气源，防止空气中的杂质产生污染。

The water and waste system supplies potable water to the lavatories and galleys and removes sink and toilet waste. The water and waste system also removes rain water from the door sill areas. The water and waste system has these subsystems: potable water, air supply, waste disposal.

The potable water system supplies water to the lavatories and galleys. The potable water system has these subsystems: passenger water, water heating, water quantity indication. The passenger water system supplies water to the lavatories and galleys. The water heating system heats the water supplied to the lavatory hot water faucets. The water quantity indication system measures and displays the quantity of water in the potable water system.

The waste disposal system removes water from the lavatory and galley sinks, human waste from the lavatory toilets, and rain water from the door sills. The waste disposal system has these subsystems: gray water, vacuum waste, waste tank quantity indication.

The gray water system has these functions: drains water and other liquids from the lavatory and galley sinks, gives exhaust ventilation from the lavatories and galleys, drains rain water from the entry/service door sill areas. The gray water system has these two subsystems: waste water system, door sill drain system. The waste water system collects the waste water from the lavatory and galley sinks and drains the waste water overboard through heated drain masts. The waste water system also gives exhaust ventilation from the lavatories and galleys. The door sill drain system collects rain water from the entry/service door sill areas and drains the water overboard through drain fittings.

The vacuum waste system removes human waste material from the toilets and holds it in the waste tank. The vacuum toilet collects human waste. A flush cycle puts the waste material into the vacuum waste tubing. Cabin pressure pushes the material to the vacuum waste tank. There are two sources that cause low pressure (vacuum) in the tank: the vacuum blower, cabin-to-ambient pressure differential. The waste is kept in the waste tank until servicing. Servicing is done at the waste service panel. The vacuum waste system has these components: toilet assembly, waste tank, liquid separator, waste tank rinse nozzle, waste tank rinse filter, waste tank rinse fitting assembly, waste drain valve assembly, waste drain ball valve, drain line blockage removal valve, vacuum check valve, vacuum blower, vacuum blower barometric switch, waste tank point level sensor, waste tank continuous level sensor, logic control module (LCM), waste quantity indicator.

The waste tank quantity indication system measures and shows the quantity of waste in the waste tank. These components are in the waste tank quantity indication system: waste tank point level sensor, waste tank continuous level sensor, logic control module (LCM), waste quantity indicator. The waste tank quantity indication system does these functions: monitors and shows the level of waste in the waste tank, stops operation of the toilets when the waste tank is full, gives indication when the lavatories stop operation for waste tank full condition, gives indication when the sensors are dirty, BIT. The water tank pressurization system pressurizes the potable water tank. Pressure for the water tank comes from the pneumatic system or the air compressor. The water tank pressurization system has these functions: controls the air pressure that goes into the water tank, selects the source of the pressurized air, prevents contamination from unwanted material in the air.

缩略语

AC　Alternating Current　交流

BIT　Built In Test　自检

C　Celsius　摄氏

CLS　Continuous Level Sensor　连续液位传感器

CONT　Control　控制

DC　Direct Current　直流

E　Empty　空

F　Fahrenheit　华氏

F　Filter　过滤器

F　Full　满

FCU　Flush Control Unit　冲洗控制装置

H　Heater　加热器

INOP　Inoperative　不工作的

IND　Indication　指示

LAV　Lavatory　厕所/盥洗室

LCM　Logic Control Module　逻辑控制模块

LED　Light Emitting Diode　发光二极管

PSID　Pounds Per Square Inch Differential　磅/平方英寸压差

PSIG　Pounds Per Square Inch, Gauge　磅/平方英寸(表压)

PLS　Point Level Sensor　点液位传感器

QTY　Quantity　量

REF　Reference　参考

S　Switch　开关

T　Transmitter　发送器

TYP　Typical　典型

专业术语

A a

按钮锁　snap-latch

奥氏体不锈钢　austenitic stainless steel

B b

泵壳体　pump housing

编织带　lacing cord

变截面节流嘴　variable throttling nozzle

标牌　placard

标签　instruction decal

冰膨胀损害　ice expansion damage

不锈钢　stainless steel

不锈钢内胆　stainless steel liner

不正确的维护　improper servicing

C c

操纵杆　actuator lever

操纵力　pilot force

厕所　toilet

厕所不工作指示灯　lavatory inoperative light/LAV INOP light

厕所冲洗泵　toilet flush pump

厕所地面服务设备　lavatory service unit

厕所通气系统　lavatory ventilation system

厕所泄放阀　lavatory drain valve
测量信号　measurement signal
层状盘式过滤器　layered disk filter
缠绕在管路上　wind around the line
超声波信号　ultrasonic signal
乘务员面板　attendant panel
乘务员位置　attendant station
冲水管　water washer tube
冲洗/注液口　flushing/charging port
冲洗阀/排放阀　flush valve
冲洗开关/冲洗按钮　flush switch
冲洗控制装置　flush control unit
冲洗控制装置测试器　flush control unit tester
冲洗控制组件　flush control unit
冲洗口　flush port
冲洗循环　flush cycle
冲洗液　flushing fluid
冲洗装置　flush equipment
抽水马桶/自循环马桶　flush toilet
抽水马桶组件/自循环马桶组件　flushing toilet assembly
臭味通气管　odor vent line
出口压力　output pressure
出水管路　water outlet line
出水口　water outlet
厨房废物处理装置　galley waste disposal unit
厨房废物处理装置冲洗阀　galley waste disposal unit flush valve
厨房废物处理装置清洗阀　galley waste disposal unit rinse valve
厨房供气管　galley air supply line

厨房供水控制阀门　galley control valve

厨房水龙头　galley faucet

传感器污染　sensor fouled/sensor dirty

磁激励器　actuator magnet

存储　storage

𝒟 𝒹

大型运输机　large transport aircraft

带式加热器　heater tapes

带状卡箍　band clamp

单向阀　check valve

挡板阀/舌型阀　flapper valve

导线外的绝缘套管　wiring loom

倒流　backflow

灯泡测试　lamp test

登机/服务门门槛区域　entry/service door sill area

地面冲洗单向阀门　ground flush check valve

地面冲洗接头　ground flush fitting

地面供气接头　ground air connection

地板排水装置　floor drain

地面通气接头　ground vent air fitting

地面维护操作　ground service operation

地面维护汇流条　Ground Service (GS) bus

地面维护设备　ground service facility

点液位传感器　point level sensor

电磁操纵的　solenoid operated

电动泵　electric pump

电动旋板泵　electric (ally-driven) vane pump

电动压气机　electric driven compressor

电加热器　electrical heater

电缆接头　electrical cable terminal splice

电气接头　electrical connector

电容　capacitance

电源开关　power switch

电源指示灯　power indicator light

垫式加温器　gasket heater

堵盖　blind cap

端盖　end closure

短S形管路构造　short “S” shaped pipe configuration

段式显示　segment display

断路开关　cut-out switch

断路器　circuit breaker

E　e

额定出口压力　rated output pressure

额定进口压力　rated inlet pressure

额定温度　rated temperature

额定压力　nominal pressure

F　f

发光二极管分段显示　LED segment display

阀门接口装置　valve interface unit

阀门位置开关　valve proximity switch

翻板式阀门　flapper valve

反虹吸阀　anti-siphon valve

反压　back pressure

负压表　vacuum gauge

方块冰　ice cube

防虹吸阀　anti-siphon valve

防虹吸管　anti-siphon tube

防护服　protective clothing

防震安装　shock mount

飞机机动和扰动　airplane maneuvers or turbulence

非金属涂层　a layer of non-stick material

废(污)水　waste water

废水排放　waste water drain

废水排放阀门/管路　waste water drain valve/line

废水排放管加热器　waster water drain mast heater

废液排放口　waste tank drain outlet

分隔板　separator

分离器盖子　separator cap

分配管路　distribution line

封头/盖子/头部　head

服务板灯　service panel light

服务板照明/读出开关　service panel light/readout switch

服务板照明灯　service panel light

服务板照明灯开关　service panel light switch

浮子　float

复位按钮　reset button

复位式过热保护开关　self-resetting thermal switch

复位压力　reseat pressure

G g

盖手柄　cap handle

盖子　cover

干涉支架　interference brackets

干线机　trunk aircraft/trunk liner

刚性导管　rigid conduit

高度压力开关接头-真空马桶　altitude pressure switch adapter-vacuum toilet

隔离玻璃　glass barrier

工作循环　duty cycle

工作压力　operating pressure

公务机/行政飞机　business/executive aircraft

供气　air supply

供气管　pneumatic supply line

供水关断阀　water supply shutoff valve

供水管　supply hoses/water supply line

供水管加温　water supple lines heat

供水及放泄管路接头　supply and drain line connection

供水截断阀　water supply shutoff valve

供水系统　water supply system

固定过滤器　fixed filter

固定螺钉　setscrew

关断活门　shut off valve

管夹　pipe clamp

管接头　tube fittings

管路接头　ducting connection

管内空气过滤器　in-line air filter

盥洗室　lavatory

盥洗室马桶　lavatory toilet

盥洗室水关断活门　lavatory water shut off valve

盥洗室水加热器　lavatory water heater

盥洗室水龙头　lavatory faucet

盥洗室洗手池　lavatory washbasin

硅胶　silicone rubber

硅胶垫　silicon liner

过滤器　filter

过滤器转接头　filter adapter

过滤元件　filter element

过热复位式开关　overheat/rest switch

过热开关　overheat switch/over temperature switch

过热控制器　overheat controller

过温　over temperature

过温保护　over temperature protection

过压活门　overpressure valve

过压开关　over pressure switch

H　h

航邮飞机　airmail aircraft

喝水水龙头　drinking fountain

恒温器　thermostat

恒温开关　thermoswitch

环形导管　ring pipe

环形加热器　annular heater

回转圆盘阀　rotating disk valve

活塞泵(柱塞泵)　piston pump

活塞式电动泵　piston motor pump

货舱后端壁　aft bulkhead of the cargo compartment

货舱内衬　cargo compartment lining

货机　cargo aircraft/airfreighter

J　j

基座　pedestal

极限操纵力　limited breakout force

疾病　illness

加热带(电热带)　heater tape (ribbon heaters)

加热的排水杆　heated drain mast
加热控制开关　heater control switch
加热控制器　heat controller
加热器开关灯　heater switch light
加热毯　heater blanket
加水/供水阀　fill/supply valve
加水/溢流阀　fill/overflow valve
加水/溢流阀手柄　fill/overflow valve handle
加水阀　fill valve
加水阀门手柄　fill valve handle
加水管接头　fill line fitting
加水管路　fill line
加水管路通气阀　fill line vent valve
加水接头　water fill fitting
加水接头及堵盖组件　fill connector and cap assembly
加水口　fill port
加水手柄　water fill handle
加水水位自动控制　water filling automatic level
加水系统　water filling system
加注程序　fill process
加注接头　fill nipple
加注孔　fill orifice
夹布胶管　diorite hose
尖锐物体　sharp object
监控循环　monitor cycle
减压阀　pressure relief valve
减压器　reducer
接近开关　proximity switch
接头　connection/adapter/coupling

接头出口　outlet fitting

接头进口　inlet fitting

节流孔　throttling orifice

截断阀/关断阀　shutoff valve

金属导管　metal conduit

金属环　grommet

金属密封圈　metal sealing ring/sealing ring

进口压力　inlet pressure

进气口　intake port

进水管路　water inlet line

径向活塞泵　radial piston pump

静止滤栅　static filter basket

聚酰胺增强编织　nomex and cres braids

巨型运输机　jumbo aircraft

绝对压力　absolute pressure

军用飞机　military aircraft

军用运输机　military transport aircraft

K　k

卡箍　clamp/clip

卡箍带　band strap

卡环　snap ring

卡式锁　snap latches

卡套　clam shell

开裂和爆裂声　cracking and popping noise

开启压力　cracking pressure

壳体　bowl

口盖　access cover

可用容积　useable volume

可用水箱容积　usable tank capacity

客货两用机　combination aircraft

空气锤　kinetic air-ram

空气过滤器　air filter

空气过滤器元件　air filter element

空气活门　air valve

空气净化器　air purifier

空气流失　escape of air

控制冲水的逻辑装置　control logic unit

控制电路　control electronic

控制杆　control rod

控制钢索　control cable

控制压力　control pressure

快卸接头　quick-disconnect fitting

快卸连接　quick-release connection

宽体飞机　wide fuselage aircraft

L l

冷气活门　pneumatic valve

冷气歧管　pneumatic manifold

冷水　cold water

冷水按钮　cold button

离心泵　centrifugal (-type) pump/radial pump

连接杆　interconnecting link

连接螺钉　attach screw/bonding stud

连续液位传感器　continuous level sensor

流量指示器　flow indicator

流水孔　water flow orifice

流向接头　flow direction arrow

六角扳手　wrench

滤网　screen

滤芯　filter cartridge

旅客供水　passenger water

旅客机　passenger aircraft

逻辑控制装置　logic control unit

M　m

马达驱动泵过滤器组件　motor-pump-filter assembly

马达驱动阀　motor operated valve

马达驱动隔断阀门组件　motor driven shutoff valve assembly

马达驱动加水管路隔断阀门　motor driven fill line shout off valve

马达组件　motor unit

马氏体不锈钢　martensitic stainless steel

马桶　toilet

马桶便盆　toilet bowl

马桶冲洗管路　bowl flush line

马桶冲洗控制开关　toilet flush control switch

马桶冲洗手柄　toilet flush handle

马桶盖　toilet cover

马桶加水/水接头　toilet fill/sluice coupling

马桶排放　toilet drain

马桶盆　toilet bowl

马桶上缘　toilet bowl upper rim

马桶手动关闭阀　toilet manual shutoff valve

马桶污物　toilet waste

马桶泄漏测试器　toilet leak tester

马桶罩　toilet shroud

马桶组件　toilet assembly

马桶座和盖　toilet seat and cover

马桶坐垫　toilet seat

毛刷　bristle brush

门槛　door sill

门槛排水　door sill drain

密封接头　seal fitting

密封圈　sealing ring

灭火剂　fire extinguishing agent

民用飞机　civil aircraft

膜盒式高度表　aneroid altimeter

膜盒压力传感器　aneroid pressure transmitter

膜盒膜片　aneroid capsule

膜片　capsule

模压热塑性内衬　molded thermoplastic liner

木炭过滤器　charcoal filter

N　n

内表面　interior surface

内部负压　negative internal pressure

内部泄漏量　internal leakage

内置过滤器　internal filter

内置过热保护器　built-in thermal overload protection

内置加热元件　integral heating elements

P　p

排放阀门控制手柄　drain valve control handle

排放盖　drain cap

排放管路　drain line

排放管去堵阀　drain line blockage removal valve

排放口　drain port

排气口　discharge port, exhaust port

排气通风　exhaust ventilation

排水　dump

排水按钮　drain button

排水阀　drain valve.

排水杆　drain mast

排水杆加热器　drain mast heater

排水杆整流罩　drain mast fairing

排放作动杆　drain rod

排水管　drain hose

排水管加温　drain line heat

排水接头　drain fitting

排水球囊　drain bladder

排水弯管　drain elbow

排水消音器　drain muffler

排水装置　drain unit

排污盖铰链　drain cap hinge

排污阀门手柄　tank drain valve handle

排污口　drain port

排污球阀　waste drain ball valve

排污系统服务板　waste system service panel

旁通阀门　bypass valve

泡沫　foam

喷射环　spray ring

喷水　spray water

喷嘴　spray nozzle

喷嘴口　spray nozzle access

偏心锁铰链　over center latch hinge

品质下降　performance deterioration

破坏压力　burst pressure

Q　q

起动压力　starting pressure/breakout pressure

气动外形　aerodynamic fairing

气动系统　pneumatic system

气滤　air filter

气滤/增益调节器　air filter/attenuator

气密　air-tight scaling

气液混合物　air liquid mixture

切断压力　cut-out pressure

清洁/校验传感器　clean/check sensor

清洗阀　rinse valve

清洗环　rinse ring

清洗接头垫圈　rinse fitting ring

球阀　ball valve

球囊　bladder an/check sensor light

球形把手　knob

驱动马达　drive motor

R　r

热保护开关　thermoswitch

热水　hot water

热水按钮　hot button

热水龙头　hot water faucets

热水器　water heater

人员产生的废水　human waste

溶剂　solvent

柔性管　flexible pipe

柔性连接软管　flexible coupling hose

软导管　flexible conduit/hose

软管夹　hose clamp

软管接头　hose connection

S s

商用飞机　commercial aircraft

商用喷气机　business jet

上电自检　power-up bit test

蛇形管　plumbers snake

设定温度　desired temperature

渗漏　seepage/leakage

实际压力　actual pressure

食品供应车　galley service truck

食品级加工油脂　food grade processing grease

释压阀门　pressure relief valve

试验夹具　test fixture

手动关断阀手柄　manual shutoff handle

梳洗柜柜门　sink cabinet door

疏通堵塞　blockage removal

疏通管　clean-out line

疏通接头　clean-out fitting

竖管　stand pipe

双重密封　dual-seal

水泵　water pump

水滴　moisture particle

水/废水　water/waste

水/废水系统　water/waste system

水传输/分配管路　water distribution line

水分离器　water separator

水管　waterline

水管及废水管防冰　water and waster lines anti-ice

水加热　water heating

水加热开关　water heater switch

水净化器　water purifier

水冷却器　water cooler

水量/水位　the quantity of water

水量表　water quantity gauge

水量变送器　water quantity transmitter

水量变送器适配电缆　water quantity transmitter adapter cable

水量指示　water quantity indication

水量指示器　water quantity indicator

水量指示系统　water quantity indicating system

水龙头　faucet

水气　moisture

水维护面板　the water service panel

水位　water level

水位传感器　water level sensor

水系统服务口盖　water service panel

水系统排放阀门　system drain valve

水系统排放手柄　system drain handle

水箱　water tank

水箱的复合结构　composite structure of the water tank

水箱供水管路　water tank supply line

水箱固定支架　water tank mounting bracket

水箱加水口　water tank fill port

水箱加压　water tank pressurization

水箱加压接头　water tank pressure connection

水箱空气加压系统　water tank air pressurization system

水箱内衬　inner lining of the water tank

水箱排放阀　water tank drain valve

水箱水位传感器　water tank level sensor

水箱通气网　water tank vent screen

水箱卸压　depressurize the water tank

水箱增压接头　water tank pressurization fitting

水箱增压系统　water tank pressurization system

水箱柱状部分/等值段　cylindrical portion of the tank

水压表　water pressure gauge/pressure meter

水银/汞　mercury

瞬时接通开关　momentarily closed switch

瞬时开关　momentary switch

四通旋转阀　four-port rotary valve

碎冰　crushed ice

碎纸　shredded paper

锁眼槽　keyhole slot

收集的水汽　trapped moisture

T　t

钛合金导管　titanium duct

弹簧负载　spring-loaded

探针　probe/wire probe

碳纤维缠绕内胆　graphite filaments wound over liner

碳纤维增强塑料容器　graphite fiber reinforced plastic container

特氟隆衬垫　teflon liner

调节螺钉　adjustment screw

通气单向阀门　vent air check valve

通气管出口　vent-pipe outlet

通气孔　vent port

通气系统　ventilation system

通用清洁剂　general purpose cleaner detergent

铜网电容传感器　copper mesh capacitance sensor

推开操纵杆　push-to-open lever

推拉杆　push-pull rod

V v

V 型卡箍　V band clamp/V-band coupling

W w

外部维护口盖　exterior service panel

外部泄漏量　external leakage

外壁　exterior wall

外表面　exterior surface

网孔盘　perforated disk

往复泵/活塞泵　reciprocating pump

微米　micron meter

微动开关　micro switch

位置开关　position switch

位置指示单向阀门　position indicating check valve

维护开关　maintenance switch

温度探针　temperature sense probe

温度调节　temperature regulation

温度选择开关　temperature selector switch

温控开关　thermal switch

微生物　microorganisms

污染　contamination

污水处理系统 waste disposal system

污水箱及过滤器清洗管路 waste tank and filter rinse line

污水排放阀组件 waste drain valve assembly

污水水量指示器 waste quantity indicator

污水维护面板 waste service panel

污水箱 waste tank

污水箱清洗管路接头 waste tank rinse line connection

污水箱清洗过滤器 waste tank rinse filter

污水箱清洗接头组件 waste tank rinse fitting assembly

污水箱清洗喷嘴 waste tank rinse nozzle

污水箱排放阀门组件 tank drain valve assembly

污水箱水量指示 waste tank quantity indication

污水箱水位传感器 toilet tank level sensor

污水箱污水液位 the level of waste in the waste tank

污物 waste material

污物清洁车 waste truck/toilet truck

污物系统 waste system

污物箱 waste tank

无纺布 lint free cloth

X x

洗涤池 wash basin

洗涤废水(灰水) grey water

洗涤盆溢流排放管 sink overflow drain line

洗涤废水排放阀 grey water drain valve

洗脸盆水龙头 washbasin faucet

洗手池 sink

细菌生长 bacteria growth/growth of bacteria

细菌和病毒 bacteria and viruses

系统最高温度　system maximum temperature

橡胶密封圈　rubber sealing ring

橡胶罩　rubber cover

消毒剂　disinfectant

消音器　noise baffle

泄漏量　leakage

卸压阀　pressure relief valve

卸压作动器　pressure relief actuator

小型储存装置　small reservoir unit

旋转喷嘴　rotating spray nozzle

循环泵　circulating pump

Y y

压差开关　differential pressure switch

压力(限制)开关　pressure limit switch

压力传感器　pressure transmitter/pressure senson

压力传感器探针　pressure transmitter probe

压力调节阀门　pressure regulating valve

压力调节器　pressure regulator

压力罐/调节器　pressure container/regulator

压力开关　pressure switch

压力指示器　pressure indicator

压气机　air compressor/compressor

压气机关断开关　air compressor cut-out switch

压气机开关　compressor switch

液面指示器传感器　liquid level indicator transmitter

液体分离器　liquid separator

液体介质　fluid medium

液体洗涤剂　liquid detergent

液位传感器　fluid level sensor

液压定时器　hydraulic timer

一次性注射器　disposable syringe

异丙醇　isopropyl alcohol

溢流/排水管路　overflow/drain line

溢流/排水口　overflow/drain port

溢流管接头　overflow line fitting

溢流及排放接头　overflow and drain connector

溢流接头(立管)　water overflow fitting (standpipe)

溢流排水管　overflow drain line

引气供应管　bleed air supply duct

引气温度　bleed air temperature

引气压力　bleed air pressure

饮用水维护面板　potable water service panel

饮用水供应车　potable water truck

饮用水加压系统　potable water pressurization system

饮用水水箱　potable water tank

饮用水系统　potable water system

饮用水循环系统　the potable water recirculation system

油脂　grease

优质耐蚀不锈钢　super-corrosion-resistant stainless steel

与外侧通气　vent overboard

雨水　rain water

预添加化学液　chemical precharge

运输机　transport aircraft

Z z

增压系统　pressurization system

增压设备　pneumatic equipment

真空 vacuum
真空泵 vacuum pump
真空单向阀 vacuum check valve
真空风扇 vacuum blower
真空风扇气压继电器 vacuum blower barometric switch
真空力 vacuum power
真空马桶 vacuum toilet
真空污水系统 vacuum waste system
支线机 regional aircraft/commuter
纸滤芯元件 paper filter element
指示 indication
煮咖啡机 coffee maker
柱塞阀 cartridge valve
柱塞阀活塞 operate cartridge valve plunger
转动装置 rotary actuator
转进转出钢索 rotary-in rotary-out cable
装饰罩 decorative cover
自动防故障装置 fail-safe device
自动零位调节 auto zero adjustment
自封接头 self-sealing coupling
自通气水龙头 self-venting faucet
紫外线 ultraviolet(uv)
紫外线灯 ultraviolet(uv)lamps
紫外线水处理装置 ultraviolet(uv) water treatment unit
中央处理系统 common core system
总管路 manifold
总容积 total volume
最大工作压力 maximum operating pressure
最内侧元件 innermost element

最小操纵力　minimum starting force

最小工作压力　minimum operating pressure

座舱与机外的压差　cabin-to-ambient pressure differential

致病微生物　disease organisms

内装饰

导 语

内装饰包含驾驶舱装饰、客舱装饰、货舱装饰及绝热层，为机组与乘员提供舒适安全的乘坐环境。内饰装由装饰板(天花板、侧壁板及门区装饰、分隔板等)与杂项设施(存储设备、遮阳板、操控台及地板覆盖物等)组成。驾驶舱装饰覆盖机头地板以上结构，多样化的设备与装饰件为机组带来舒适、便捷与安全的生活、工作环境。客舱装饰分布在机身地板上，座舱被分为公共区域、旅客乘坐区，包含机组旅客门及应急出口，公共区域临近座舱入口。货舱装饰用于保证货物的控制与安全。绝热层安装在气密舱内，作用为减少机体热量流失，防止冷凝水形成，以及减少座舱噪声。

The equipment and furnishings includes cockpit, passenger compartment and cargo compartment equipment and furnishings and thermal insulation which are installed in the aircraft give comfort and safety to passengers and crew. The equipment and furnishings comprises of interior panels(ceilings, sidewall and doors panels, partitions etc.) and miscellaneous(towages, sunvisors, window blinds, consoles, floor covering etc.). The cockpit furnishings comprise the area above the floor structure. Various furnishings and equipment are fitted in the cockpit for the comfort, convenience and safety of the occupants. The passenger compartment is in the upper fuselage section. The cabin is divided into utility areas and seating areas. The passenger/crew doors and emergency exits are also included in the cabin area. The utility areas are adjacent to the cabin entrances. Partitions and curtains divide the utility areas from the seating areas. The equipment and furnishings are installed in the cargo compartments for handling and safety of the cargo. The thermal insulation is

installed in the pressurized cabins, minimizing the loss of heat from the fuselage, stops the formation of condensation, and reduces the noise level in the fuselage.

缩略语

ACTD　Advanced Concept Technology Demonstration　先进概念技术演示验证

AFM　Aircraft Flight Manual　飞机飞行手册

AMM　Aircraft Maintenance Manual　飞机维修手册

AMS　Aeronautical Material Specification　航空材料规范

BHD　bulkhead　隔框

BL　Buttock Line　纵剖线

CL　Center Line　中心线

CMR Certification Maintenance Requirements　审定维护要求

CSP　Certification support plan　符合性验证支持计划

FICD　Functional Interface Control Document　功能接口控制文件

GSE　Ground Support Equipment　地面保障设备

LOI　Letter of Intent　意向书

MCD　Movable Class Divider　可移动分舱板

MICD　Mechanical Interface Control Document　机械接口控制文件

MIL-STD　Military Standard　美军标

MMEL　Master Minimum Equipment List　主最低设备清单

MP　Maintenance Program　维修大纲

MPP　Maintainability Program Plan　维修项目计划

MRBD　Maintenance Review Board Document　维修审查委员会文件

MSG　Maintenance Steering Group　维修指导小组

MTA　Maintenance Task Analysis　维修任务分析

MTBF　Mean Time Between Failures　平均故障间隔时间

MTTR　Mean Time To Repair　平均修复时间

OHSC　Overhead Stowage Compartment　顶部行李箱

P/N　Part Number　零件图号

PC　Polycarbonate　聚碳酸酯

PDPD　Preliminary Design and Performance Declare　初步设计与性能声明

PIU　Passenger Information Units　旅客信息装置

PQPP　Preliminary Qualification Program Plan　初步鉴定项目计划

PSS　Passenger Service System　乘客服务系统

PSSA　Primary System Safety Assessment　初步系统安全性评估

PSU　Passenger Service Unit　旅客服务单元

QPP　Qualification Program Plan　鉴定项目计划

QTP　Qualification Test Procedures　鉴定试验程序

RFP　Request for Proposal　邀标书

SSA　System Safety Assessment　系统安全性评估

TC　Type Certificate　型号合格证

TSO　Technical Standard Orders　技术标准要求

专业术语

A a

A检　A check

A向视图　A view

安全带　safety harness/safety belt

安全性　safety

安装　installation/mount

安装尺寸　mounting dimension

安装接头　attachment fitting

安装连杆　mount link/rod

安装图　installation drawing

安装支承架　support bracket

暗适应　dark adaptation

凹痕　dent

凹陷型天花板　recessed ceiling

B b

白　white

百分数　percent

钣金工艺　sheet forming

钣金工艺规程　sheet metal process planning

钣金下料　blanking/sheet cutting

半径　radius

半物理仿真　semi/full-physical simulation

半圆头螺钉　half-round screw

半圆头铆钉　brazier head rivet/snap head rivet/half-round head rivet

磅　pound

薄垫片　shim

保养　servicing

保障性　supportability

报告　reports

备件　wear-out parts/replacement part

备用冗余　standby redundancy

倍数　times

苯　benzene

壁橱　closet

壁挂温度计　wall thermometer

扁圆头铆钉　button-head rivet/brazier head rivet

变量　variables

变形　deformation

标称直径　nominal diameter

标牌　identification plate

标签　label/tag

标题栏　index/title panel

标准件　standard parts

表面处理　surface treatment

表面缺陷　blemish

表面损坏　superficial damage

丙烯　propylene

玻璃　glass

玻璃钢　fiberglass-reinforced plastic

玻璃丝　glass wool

玻璃纤维布　fiberglass cloth

箔片　foil

泊松比 Poisson ratio

补偿 compensation/offset

补丁 patches

不定期检查 casual inspection

不定期维护(非计划性维修) unscheduled maintenance

不合格品 nonconforming product

不可修复产品 non-repairable item

不透明塑料 opaque plastic

不锈钢 stainless steel/rustless steel/noncorrosive steel

部件 components

C c

材料标准 materials standards

材料特性和说明 materials attribute and introductions

彩度 chroma

参考量 reference

舱门 doors

舱门过道侧壁板 doorway sidewall panel

舱门过道侧壁衬固定 doorway sidewall lining mount

操纵台 console/control pedestal/control stand/control panel

槽钢 channel steel

草图 sketch

侧 lateral/side

侧壁板 sidewall panel

侧壁板上遮光帘 blinds on the sidewall panel

侧壁手册柜 sidewall manual stowage/closet

侧壁下盖板 sidewall lower closure

侧壁装饰 sidewall furnishing

侧衬板 side lining panel

侧视图　side view drawing

测试技术　technique of measurement and test

测试性　testability

层间剪切强度　interlaminar shear strength

层压成形　laminating moulding

插孔板　jack panel

插头　receptacle plug

插座　socket

茶杯座　cup holder

差　difference

拆开检验　strip inspection

拆卸　removal/dismount

产品生产图样　product's production patterns

产品支援工作计划　product support program plan

产品状态　item state

长度　length

长方形　rectangle

长桁　longitudinal stringer

常用消耗性材料　common consumptive material

超过　more than/exceed

超声　ultrasound

超声速飞机　supersonic aircraft

超重　overweight

车削　turning

沉头铆接　countersunk head riveting

衬套　bushing

撑杆　support link/rod

成形工艺　forming technology

承力结构　carrying structure

乘号 multiplication sign

乘客舱 passenger compartment/cabin

乘客门 passenger door

乘客娱乐系统 passenger entertainment system

乘务员呼叫灯 attendant call light

橙 orange

持续适航 continued airworthiness

尺寸 dimensions/size

尺寸链图 dimensional-chain chart

尺寸图 dimensioned drawing

尺寸线 dimension line

尺码 size

齿形板 serrated plate/lamina denticulate

冲击 shock

冲击脉冲 shock pulse

冲击试验 impact test

冲孔/穿孔 punching/perforation

冲孔 punching/piercing

冲洗 flush

冲压 stamp

重复检查间隔 repeat inspection interval

抽查 random inspection

初始推荐备件清单 initial recommend spare parts list

除号 division sign

除油 degrease

厨房 galley/buffet

触觉 tactile sensation

窗 window

窗玻璃 windowpane

窗固定接头　window retainer fitting

窗轨(半)槽　window track reveal

窗槛　windowsill

窗框半边槽　window frame side slot

窗帘　window curtain

窗台　sill

窗遮光板　window shade

窗周缘密封件　peripheral seal

垂直线　vertical line

纯铜(紫铜)　copper

醇　alcohol

次供应商选择和管理计划　sub-suppliers selection and management plan

次声　infrasound

次要级　minor

刺激　stimulus

粗实线　solid line/heavy line

淬火　quenching/hardening

存物箱　stowage bins

锉修　file out

错觉　illusion

D　d

搭接　overlap joint

搭接线　bonding strap/bonding jumper

搭铁线　bonding jumper, ground connector

打保险　lock wire

打补丁　patch up

打样图　proof pattern/drawing sampling/layout drawing

大型运输机　large transport aircraft

代料单　material substitution sheet

带钢　winding steel/strip steel

单层隔层　monolithic/single layer insulation

单过道　single aisle

单面铆接/盲铆　blind riveting

单通道撤离滑梯　single channel evacuation slide

淡蓝　light blue

淡绿　light green

氮化　nitriding

挡风屏　window screen/windshield

刀具　cutting tools

导管　pipe/tube/duct/line

导线　wire

倒角　chamfering

倒圆　radiusing

灯架/座　lamp holder

灯屏　light screen

灯罩　lamp cover

灯罩面板　lamp cover panels

登机门　entry door

登机梯门　airstairs door

等于或大于　equal to or greater than

等于或小于　equal to or less than

低合金钢　low alloyed steel

滴　dripping

滴水防护件　drip-shield

底部　bottom

底漆　primer/priming paint/prime coat/priming lacquer

底视图　bottom view drawing

底图　original/vellum/base map

地板　floor/deck

地板安装式贮存组件　floor mounted storage component

地板覆盖层　floor covering

地板梁　floor beam

地板密封　floor sealing

地板装饰　floor furnishing

地面共振试验　ground resonance test

地面检查　ground check

地面维护　ground maintenance

地勤人员　ground crew

地毯　carpet

点画线　dotted line/dash dot line

电插头　electrical plug

电磁干扰和电磁兼容性　electromagnetic interference and compatibility

电镀　plating

电镀合金　alloy plating

电化学腐蚀　electrochemical corrosion

电话机托架　handset holder

电缆　electrical cable

电缆敷设图　cabling diagram

电连接器　electrical connector

电气特性　electrical characteristic

电线　electric wire

电源和照明系统　power supply and lighting system

电子干扰飞机　electronic counter-measures aircraft

垫片/圈　gasket/washer

雕塑天花板支架　sculptured ceiling bracket

吊杆　suspension link/rod

丁烯 butylene

顶部 top

顶角 apex angle

定量分析 quantitative analysis

定期检查 scheduled inspection/periodic inspection

定期试验台检查 periodic bench check

定期维护 routine maintenance

定位销 positioning pin/locating pin/fitting pin/jaw/adjustable pin/adjusting pin/set pin

定性分析 qualitative analysis

动机 motivation

动强度 dynamic strength

动态人体参数测量 dynamic anthropometric parameters

堵帽 blanking caps

镀镉 cadmium plating

镀铬 chromium plating

镀锌 galvanization/galvanize

断开 disconnect/detach

断路器板 circuit breaker panel

锻件图 forging drawing

锻造工艺 forging technology

对角线 diagonal line

对接 butt joint

钝角三角形 obtuse triangle

多用途灯 utility light

多用途飞机 all-purpose aircraft

E e

耳机 headphone/headset

二次成型　post forming/secondary forming

F f

法兰　flange

翻新　refurbishing

翻修　overhaul

翻修时限/间隔　overhaul intervals/time between overhaul

翻修寿命　overhaul life

反恐　anti-terrorist

反射　reflex

反时钟方向　counterclockwise

反应　response

返工　rework

方案论证报告　scheme demonstration report

方向　direction

防爆性　explosion-proofness

防滑　slip-resistant

防火　fire-resistant

防火系统　fire protection system

防蚀漆　anticorrosive paint

防水性　water proofness

防松螺母　blocking nut/jam nut/locking nut/pinch nut

防锈漆　anti-rusting paint/rust-resistant paint

防烟眼镜盒　smoke goggles box

飞机寿命　aircraft life

飞机维护/维修　aircraft maintenance

飞机系统　aircraft systems

飞行舱　flight compartment/deck

飞行操纵系统　flight control system

飞行后检查　postflight check

飞行后维护　postflight maintenance

飞行模拟试验　flight simulating test

飞行起落次数　number of flight cycles

飞行前检查　preflight check

飞行前维护　preflight maintenance

飞行小时　flight hours

飞行员舱门　crew door/flight station door/control cabin door

飞行员检查单　pilot's check list

非定期检查　unscheduled inspection

非金属材料　non-metallic material

非金属工艺　non-metallic material processing

非现场维修　off-site maintenance

废品　discard

废物桶　waster container

分贝　decibel

分隔软管　section hose

分解　disassemble

分解图　exploded view

分米　decimetre

分系统　subsystem

酚芳族聚酰胺　aramid phenolic

酚醛　phenol

酚醛树脂　phenolic resin

粉红　pink

风挡　windshield

风险分析和消除计划　risk analysis and abatement plan

蜂窝夹层结构　honeycomb sandwich structure

蜂窝夹层结构胶粘剂　adhesive for honeycomb-sandwich structure

蜂窝夹心　honeycomb core

蜂窝结构工艺　honeycomb structure manufacturing processes

缝纫工艺规程　sewing process procedure

扶手　grip rails/handrail

服务舱门　service door

服务员面板　attendants panel

符合性　conformability

俯视图　top view drawing

辅助表图样　supplement drawing

辅助结构　secondary structure/auxiliary structure

辅助设备　ancillary equipment

腐蚀　corrosion/erosion

附注　note

复合材料　composite material

复合材料结构　composite material structure

复合层压板　composite laminates

复合铝板　clad aluminium

复位按键　reset push buttons

副驾驶操纵台　the co-pilot console

副驾驶员　co-pilot/first officer

副连杆　auxiliary link/rod

副总工程师　vice chief engineer

副总经理　vice president

腹板　web

G　g

改型飞机　modified aircraft

干涉铆接　interference-fit riveting

干线机　trunk aircraft/trunkliner

刚度　rigidity/stiffness

钢板　steel sheet

钢化玻璃　toughened glass/heat-strengthened glass pre-straining glass

钢索/线缆　cable

高度　height

高级复合材料成形工艺　advanced composites fabrication technique

高级客舱　lounge

高效能分析　high efficiency analysis

搁腿架　leg rest

隔板　bulkhead/partition

隔框　frame

隔离器　isolator

隔热隔声层　thermal and acoustic insulation

隔振器　vibration isolator

个人通风出口　individual ventilation outlet

个人通风软管　gasper air hose

个体防护装置　personal protective units

更改单　change notice

更换　replace/change

工程塑料　engineering plastics

工具箱存放处　flight kit stowage/the toolbox repository

工艺　technology/process

工艺方法/过程　technological process

工艺分离面　production breakdown interface

工艺规程/工艺路线　process procedure

工艺规范　process specification

工艺流程　technology procedure

工艺流程图　flow diagram

工艺路线　process routine

工艺文件　process/technology documents

工艺性　technology features

工艺装备　tooling

工艺装备清单　tooling list

工艺装备图样　tooling drawing

工字钢　I beam

工作负荷　workload

工作区　workplace

工作冗余　active redundancy

公差　tolerance

公共区　utility area

公斤　kilogram

公务机/行政勤务飞机　business/executive aircraft/corporate aircraft

公务喷气机　business jet

功能试验　functional test

功效区　ergonomic margin

供应商　supplier

共振频率　resonance frequency

钩环　shackle/hook

构型管理计划　configuration management plan

构型控制程序　configuration control process

构造图　constructional drawing

古铜色　bronze

固定带　restraint strap

固定杆　holding link/rod/fixed link

固定隔板　permanent bulkhead

固定式窗　fixed window

固化　cure

固有可靠性值　inherent reliability value

固有能力　capability

固有频率　natural frequency

固有振动特性　natural vibration characteristics

故障　failure/fault

故障产品　faulty item

故障定位　fault-localization

故障分析　troubleshooting failure analysis

故障隔离　fault isolation

故障率　failure rate

故障树　fault tree

故障原因　failure cause

观察窗　optical viewer/prism window

观察飞行员　aircraft observer

管夹　pipe clamp/pipe clip

管接头　pipe union/tube union/union joint/union coupling

硅　silicon

硅橡胶　silicone rubber

轨道　rail/track

过道灯　aisle light

过渡线　transition line

过盈连接　interference fitting

H　h

焊接　welding

焊接工艺　welding technology

焊料　solder

航空材料　aeronautical material

航空工艺　aeronautical technology

航线检查　line check

航线维护人员 airline crew

航邮飞机 airmail aircraft

毫米 millimeter

合成蜂窝材料 synthetic honeycomb material

合成塑料 synthetic plastics

合金钢 alloy steel

褐/棕 brown

赫兹 hertz

黑色金属材料 ferrous material

桁条 stringer

横道 cross aisle

后部 back

后固化 post cure

后货舱 after cargo compartment

后机身/机身后段 after fuselage/rear fuselage

后视图 rear view drawing

厚度 thickness

弧线 arc line

互换性 interchangeability

护罩 shroud

滑动窗遮光帘 sliding window roller-blind

滑动桌 sliding-table

滑轮 pulley

化学腐蚀 chemical corrosion

环/圈 ring/collar

环境条件 environmental conditions

环境振动试验 environmental vibration test

环氧酚醛树脂 epoxy phenolic resin

环氧树脂 epoxy resin/epoxide resin

黄铜　brass

恢复　restoration/recovery

回程起飞前检查　turnaround inspection

回火　tempering

回转扶梯　circular stairway

活塞式飞机　piston engine airplane

货舱侧壁板　cargo compartment sidewall panel

货舱地板　cargo compartment floor panel/deck

货舱端壁　cargo compartment end wall

货舱绝热层　cargo compartment insulation

货舱门　cargo door

货舱门槛　door sill/threshold

货舱天花板　cargo compartment ceiling

货舱装饰　cargo compartment furnishing

J j

机加工艺　machining process

机加设备　machining equipments

机库内维护　hangar maintenance

机内广播系统　passenger address system

机内通话系统　interphone system

机内照明　interior/internal lighting

机上维护　on-board maintenance

机身横截面　fuselage cross section

机身蒙皮　fuselage skin

机身上部　upper fuselage

机身下部　lower fuselage

机身中上部　mid-upper fuselage

机体　airframe/body

机务人员/维修人员　maintenance personnel
机械冲击　mechanical shock
机械加工　machining
机械加工工艺规程　process procedure for machining
机械特性　mechanical characteristic
机械振动　mechanical vibration
机翼面积　wing area
机载设备鉴定报告　equipment qualification report
机组资料/报告盒　crew information bulletin holder
积　product
基本结构　basic structure
基本型飞机　basic aircraft
基础代谢　basal metabolism
基地维护　base maintenance
基体　matrix
基准　reference/datum
基准面　reference surface
基准偏差　datum drift
基准水平面　datum water level
基准线　reference line/datum line
极限载荷　ultimate loads
集装箱货舱　container cargo compartment
几何参数　geometry parameter
挤出成形　extrusion moulding
挤压　squeezing/extrusion
挤压强度　crushing strength
计算机仿真　computer simulation
计算机辅助工程　computer aided engineering
计算文件　calculation documents

记忆　memory

技术数据　specification data

技术员　technician

季检查　quarterly check

加工余量　machining allowance

加号　plus sign

加宽机身　widened fuselage

加强桁条　heavy stringer

加强结构　reinforced structure

加速振动试验　accelerated vibration test

加压头盔　pressure helmet

夹层结构　sandwich structure

夹具　fixtures

甲醇　methanol

驾驶舱　flight compartment/cockpit/flight station/flight deck

驾驶舱机组人员　cockpit crew

驾驶舱设备中心　flight compartment equipment center

驾驶舱设施　cockpit facilities

驾驶舱照明　cockpit lighting

间隔件　spacer

间隙　clearance

兼容性　compatibility

减号　minus sign

减震垫圈　damping washer/shake-proof washer

减震器　snubber/damper

减震支架　shock mount

剪切强度　shear strength

检测方法　test method

检查　inspection check

检查/功能检查　inspection/functional check

检查/观察窗　inspection/observation windows

检查项目单盒　check list holder

检修门　access door

检验/检查　inspect/check

简图/方块图　block drawing

舰载飞机　shipboard aircraft

鉴定试验　qualification test

鉴定试验报告　qualification test report

键连接　key joint

浇铸成形　casting moulding

胶接接头　adhesive-bonding joint/glued joint

胶结蜂窝结构　honeycomb bonded structure

胶粘剂　adhesive

角钢　angle steel

角形桁条　angle stringer

脚踏　footrest

铰孔　reaming

铰链　hinge

教练机　training aircraft

接头　connection/fitting/joint/adapter/coupling

接头系数　joint efficiency

节　knot

结构钢　structural steel

结构强度　structural strength

结构图　structure drawing

截面　section

金色　golden

金属标牌　metal placard

金属箔　metal foil

金属材料　metallic material

金属腐蚀　metal corrosion

金属腐蚀及防护工艺　metal corrosion and preventive technology

紧固件　fastener

紧急出口　emergency exit

进场着陆图夹　approach plate holder

进气格栅　air inlet grill

禁止吸烟与系紧安全带标识　no smoking/fasten seat belt sign

经度　longitude

经济舱　economic class

静强度　static strength

静态人体参数测量　static anthropometry

救生筏舱　life raft compartment

救援机　search and rescue aircraft

局部破坏　local failure

巨型运输机　jumbo aircraft

锯齿形板　serrated plate

聚氨酯泡沫　urethane foam

聚苯乙烯　polystyrene

聚合剂　polymerize

聚氯乙烯　polyvinyl chloride

聚醚酰亚胺　Ultem

聚四氟乙烯　polytetrafluoroethylene

聚乙烯　polyethylene

聚乙烯塑料　polyethylene plastics

聚酯薄膜　mylar

绝热层固定/安装　isolation mount

绝缘材料　insulating material

军用飞机　military aircraft

K　k

卡爪　pawl

开关　switch

开口垫圈　open washer/split washer

开门触发器　striker

开箱检验　inspection

抗剪螺栓　shear bolt

抗拉螺栓　tension bolt

抗震性　vibration strength

壳　shell

可达性措施　accessibility provisions

可靠性分配　reliability allocation/apportionment

可靠性工作计划　reliability program plan

可靠性计算　reliability accounting

可靠性框图　reliability block diagram

可靠性模型　reliability model

可靠性预计和分配报告　reliability prediction & allocation report

可燃性　flammability

可收回　retractable

可卸隔板　removal bulkhead

可卸天花板导轨　removable ceiling track

可卸天花板机构　removable ceiling mechanism

可信性　dependability

可行性　feasibility

可行性论证文件　feasibility demonstration document

可修复产品　repairable item

可用性/有效性　availability

克　gram

客舱侧壁板　cabin sidewall panel/passenger compartment sidewall panel

客舱乘务员　cabin attendant/crew

客舱分隔板　class divider partition/class divider

客舱分隔帘　class divider curtain

客舱过道　passenger aisle

客舱门侧壁板　passenger door sidewall panel

客舱舷窗装饰环　decor-ring

客舱照明　passenger cabin lighting

客货两用机　combination aircraft

空调系统　air conditioning system

空气分配系统　air distribution system

空勤组　aircrew/flight crew/aircraft crew

空勤组备用柜　crew alternate closet

空心铆接　pop rivet

空中加油机　air refueling aircraft

孔径　bore diameter

跨声速飞机　transonic aircraft

快拆紧固件　quick-release fasteners

快速固化　expediting setting

快卸隔板　quick-release bulkhead

宽度　width

宽体飞机　wide fuselage aircraft

捆包　wrap

扩口　flaring

扩埋头窝　countersinking

扩音器　megaphones

扩展　propagation

括号　brackets

ℒ *l*

拉杆　pull link/rod

拉挤成型　pultrusion moulding

拉伸强度 tensile strength

蓝图　blueprint/drawing

拦阻　retention

牢固性　security

老化　aging

类型　type

累积　accumulation

冷工艺　cold-technology

冷凝水密封件　condensate seal

冷气系统　pneumatic system

冷作硬化　cold work hardening

厘米　centimeter

离位检查　off-aircraft check

理论图　theory drawing

立方　cube/to the third power

立方体　cube

连杆　tie rod

连接杆　interconnecting link/rod

连接强度　joint strength

梁　beam/spar

亮度对比　luminance contrast

裂缝　crack

临近　adjacent

菱形　rhombus

领航员　pilot-navigator

硫化橡胶　vulcanized rubber/perduren

六角扁螺母　hexagon thin nut

六角槽形螺母　hexagon castle nut/hexagon slotted nut

六角防松螺母　hexagon check nut/hexagon lock nut

六角头螺栓　hexagon head bolt/hex head bolt

六角自锁螺母　self-locking hexagon nut

旅客乘坐区　passenger seated area

铝合金　aluminum alloy

铝镁合金　aluminum-magnesium alloy

绿　green

轮廓线　contour line

螺钉　screw

螺接　screw joining

螺母　nut

螺栓　bolt

螺旋桨　propeller

M　m

埋头/沉头螺栓　flush bolt/countersunk bolt/flush-headed bolt

埋头铆钉　countersunk head rivet/sunk head rivet

毛坯图　block drawing(blank drawing)

铆钉　rivet

霉变　mildewing

门的衬板　door lining

门顶标志　overdoor sign

门顶灯　overdoor light

门机构　door mechanism

门结构　door structure

门槛　doorsill/door threshold

门口保护网　door protector net

门框　door frame/door-case

门框装饰　doorcase decoration

门梁　door beam

门区顶部装饰板　doorway header panel

门区周围装饰　door surround

门手柄/把手　door handle/knob

门锁　door lock

门锁闩　door latch

锰　manganese

米　meter

密封垫圈　seal washer/joint washer/tightening seal washer

密封剂　sealant

密封框　pressure bulkhead

密封条　sealing strips

面积　area

面罩袋　mask container

灭火器　fire extinguisher

灭火系统　fire extinguishing system

民航飞行员　airline pilot/civilian pilot

民用飞机　civil aircraft

敏感性　sensibility

明度　lightness

明细表　specified list

模具　die/mould

模型图　model drawing/diagrammatic figure

磨料　abrasive

磨削　grinding

目录　table of contents

N n

内部　interior

内部的　inboard/internal/inside

内层玻璃　inner pane

内窗玻璃　inner window pane

内径　inside diameter/inner diameter

耐火漆　fire-proof paint

耐久试验　endurance test

耐久性　durability

耐久性试验　endurance test

耐磨漆　abrasion resistant paint

耐振试验　endurance vibration test

瑙加海德人造革（商标）　naugahyde

尼龙　nylon

腻子　putty

啮合　engage

啮合接/榫接　joggle joint

拧紧　screw up/tighten

拧松　unscrew/screw off/loosen

牛鼻板　bull nose

P p

爬升　climb

排气格栅　air exhaust grill

盘头铆钉　panhead rivet

抛光　polishing

抛物线　parabola

刨削　planing shaping

配合性　compliancy
喷丸强化　shot peening
碰撞　bump
批生产飞机　mass-produced aircraft
疲劳和断裂强度　fatigue and fracture breaking strength
疲劳极限　fatigue limit
偏心弹簧卡销　overcenter spring detent
拼接　joggle
频道选择器　channel selector
频率　frequency
平板　plate
平垫圈　flat washer/plain washer
平方　square/to the second power
平接头　flush fittings
平均使用期限　mean service life
平面　plane
平面图　plane view drawing
平头螺钉　flat end screw，flat screw
平头螺栓　flat head bolt
平头铆钉　flat head rivet/flash rivet
平行四边形　parallelogram
平行线　parallel line
坡度/斜率　slope
破坏试验　destruction test
剖面图　section view drawing
剖面线　section line
剖视图　cut-away view drawing

Q q

漆/涂料　paint/coating

启封　unsealed

起飞前最后检查　preflight last minute check

气密单耳托板自锁螺母　airtight self-locking plate nut with one lug

铅灰色　leaden

前货舱　forward cargo compartment

前机身/机身前段　forward fuselage/front fuselage

前面　front

前视图/主视图　front view drawing

嵌接　insert joining

强度分析报告　strength analysis report

强度极限　ultimate strength

强度计算　structure calculation

强化玻璃　tempered glass

墙面板　wall covering

墙裙/踢脚板装饰板　dado panel

切边　edge cutting

切口　slot cutting

青铜　bronze

清理　clean out

清漆　varnish

清洗　cleaning

情绪　emotion

球面　spherical

球体　spheroid

区　zone

区域检查　zonal inspection

曲线　curved line/line graph

屈服比　yield ratio

屈服极限　yield limit

躯干腿长指数　index of the trunk and leg

去尖角　remove sharp/break sharp edges

去毛刺　deburr/burring

全计算仿真　full-mathematic simulation

权衡研究　trade study

缺陷　defect

ℛ　r

燃气涡轮式飞机　turbine-engine aircraft

燃油系统　fuel system

热处理　heat treating

热处理工艺　heat treatment technology

热处理工艺规程　process procedure for heat-treatment

热工艺　metallurgical technology

热塑料　thermoplastic

人机功效　ergonomics

人机界面　human machine interface

人际关系　interpersonal relation

人能力　capacity

人失误　human error

人体测量　anthropometry

人体模型　human model

人体质　constitution

人体坐标　human coordinate

人先天素质　diathesis

人效能　human performance

人造革　imitation leather

认知　cognition

任务可靠性　mission reliability

任务剖面　mission profile/envelope
韧性　toughness
日常检查/例行检查　routine check
日历年　calendar year
绒织部分　flocked sections
容差分配　tolerance/allowance allocation
容量　capacity
蠕变　creep
入口照明灯　entryway light
软件　software
软质皮革　chamois leather/soft leather
锐角三角形　acute-angle triangle
润滑/保养　lubrication/servicing
润滑剂　lubricant

S s

塞子　plug
三角形　triangle
三棱柱　triangular prism
三面图　three-view drawing
三喷气发动机飞机　tri-jet aircraft
散装货舱　bulk cargo compartment
色调　hue
砂纸　sandpaper
闪烁　flicker
扇形　sector
商　quotient
商务舱　business class
商用飞机　commercial aircraft

上部 upper

上层舱地板 upper deck

上层客舱 upper deck passenger compartment

上偏差 above nominal size/upper deviation

设备/装饰 equipments and furnishing

设备 facilities

设备架 equipment racks

设备图 equipment drawing

设计报告 design report

设计补偿 design offset

设计定型图样 finalized design drawing

设计方案图 design scheme drawing

设计分离面 initial breakdown

设计规范 design specification

设计基准 design datum

设计计算和分析报告 design calculate and analysis report

设计任务书 design task paper

设计手册 design manual

设计图 design drawing

设计要求 design requirement

设计约束条件 constraint

设计载荷试验 ultimate load test

设计重量 design weight

身长坐高指数 index of height and trunk

身体包络面 body envelope

深度 depth

深红 crimson

深蓝 dark blue

深绿 dark green

升　liter

升压塑化　step-up cure

生产定型　production finalization

生理耐受性　physiological tolerance

生理稳态　physiological steadiness

生物反馈　biofeedback

生物节律　biorhythm

声压级　sound pressure level

声振试验　acoustic environments test/acoustic vibration test

剩余强度系数　coefficient of residual strength

剩余寿命　remaining life

失效原因/机理　failure cause/mechanism

失重　weightlessness

十字槽螺钉　cross recessed screw/phillips screw

十字槽埋头自攻螺钉　cross recessed countersunk tapping screw

时限　time limits

实线　actual line

实心铆钉　solid rivet

实验室试验　laboratory test

使用/目视检查　operational/visual check

使用检查　operational check

使用可靠性值　operational reliability value

使用寿命　service life/operating life/useful life

使用载荷试验　limit load test

示意图　diagrammatic drawing

视觉　vision

视敏度　visual acuity

视野　visual field

试飞　flight-test

试飞备件清单　flight test spare parts list
试飞员　flight test pilot
试验/试飞工艺规程　process procedure for test and flight test
试验程序　test means of system
试验件　specimen
试验件图　test piece drawing
试验项目　test items of system
试装　trial assembly
适航性　airworthiness
适用性　applicability
释压　relief pressure
释压口盖　blowout panel
手持麦克　handheld mic
手提式灭火器　portable fire extinguisher
手提箱固定器　suitcase restraint
首次翻修前平均寿命　mean life to first overhaul
首次检查期　inspection threshold
首翻期　overhaul threshold/time to first overhaul
寿命单位　life unit
寿命剖面　life profile/envelope
书报架　magazine rack
书写台　writing table
舒适区　comfort margin
树脂　resin
双层玻璃窗　double-glazed window
双层舱机身　two-deck fuselage
双耳自锁螺母　self-locking nut with two lugs
双过道　double aisle
双喷气发动机飞机　twin-jet aircraft

双曲线　hyperbola

双通道客舱　two-aisle cabin

水/污物　water/waster

水陆两用机　amphibian

水密性　water tightness

水平线　water line/horizontal line/level line

顺时针方向　clockwise

四边形　quadrilateral

四喷气发动机飞机　four-jet aircraft

塑料　plastics

塑性　plasticity

随机工具　tools going along with aircraft

损伤　damage

T t

钛钢　titanium steel

钛合金　titanium alloy

泰德拉膜　Tadlar foil

探测器控制装置　detector control unit

探测系统　detection system

碳素钢　carbon steel

逃生绳　escape rope

逃逸滑梯　escape slide

弹簧　spring

弹簧垫圈　spring washer/retaining washer/grower washer

弹簧钩　spring hook

弹簧卡箍/卡子/夹　spring clamp/spring clip

弹性极限　elastic limit

套筒/套管　sleeve

套筒螺母　turnbuckle

特点/性　feature/characteristic

特殊详细检查　special detailed inspection

特种加工工艺　non-conventional machining

特种液体　special fluid

梯度　gradient

提升连杆　lift link/rod

体段　segment

体型　somatotype

天花板检修口盖　ceiling access panel

天花板框架　ceiling frame

天花板直接照明灯　direct ceiling light

天蓝　azure/sky blue

填充条　filler strips

条件反射　conditioned reflex

调整　adjustment

调整杆　adjusting link/rod

调准　alignment

贴花层　decal

铁　iron

听觉　auditory sensation

听力损失　hearing loss

听阈　auditory threshold

通电检查　power-on check

通气孔　vent hole

通信系统　communication system

通用插座　utility socket

通用工具　universal tools

同步连杆　synchronizing link/rod

头等舱　first class

投影图　projection drawing

透明材料　transparent material

透视图　perspective drawing

凸起型天花板　bullnose panel

图区　zone

图样　drawing

涂层/镀层　coating

涂胶量　glue-spread

腿部空间　leg room

退火　annealing

托板螺母　anchor nut/plate nut

椭圆　ellipse

W w

外部的　outboard external/outside

外部监控检查　external surveillance inspection

外层玻璃　outer pane

外场维护　line maintenance/airfield maintenance

外敷纤维层的分隔网　fabric covered divider nets

外径　outside diameter/outer diameter

外廊尺寸　overall dimensions, outline dimension

外形　appearance, contour

外形图　contour drawing/outside view

外展　abduction

外罩组件　shell assembly

弯度　camber

弯曲　bending

完整性　integrity/completeness

微米　micrometer
危险级　hazardous
维护检查　maintenance check
维修/维护　maintenance
修理　repair
维修分离面　maintenance break
维修工作类型　maintenance task type
维修间隔单位　maintenance interval unit
维修时限/间隔　maintenance intervals
维修实施　maintenance practices
维修性　maintainability
维修性分配　maintainability allocation (apportionment)
维修性分析报告　maintainability analysis report
维修性工作计划　maintainability program plan
维修性模型　maintainability model
维修性预计　maintainability prediction
尾端　end
纬度　latitude
卫生间/盥洗室　lavatory
位置　position
稳定性　stability
涡轮风扇式飞机　turbofan aircraft
涡轮螺旋桨式飞机　turbo-prop aircraft
涡轮喷气式飞机　turbojet airplane
卧铺　berth
无光　matt
无损检测　nondestructive testing
无头铆钉　slug rivet/headless rivet
无图件号　dash number

物理特性　physical characteristic

雾度　haze

X　x

铣削　milling

系统可靠性分析报告　system reliability analysis report

系统设计符合性报告　system design compliance report

系统效能　system effectiveness

细实线　fine line

细纹　grain

下部　lower

下部天花板架　lower ceiling support

下降　descent

下偏差　below nominal size

弦线　chord line

现场发泡　foam in place

现场维修　on-site maintenance

线　line

限止件　stopper

限制载荷　limit loads

相交线　intersecting line

镶接/拼接　splicing

镶进　insert into

镶嵌件　insert

详图/零件图　detail/part drawing

向后　afterward

向前　forward

向上　upward

向下　downward

向右　rightward

向左　leftward

橡胶　rubber

橡胶模压密封件　rubber molded seal

消耗件/品清单　consumable parts/items list

销钉　pin

销连接　pin joining

小数点　decimal point

校正　correction

校准　calibration

楔形　wedge

斜侧壁　sloping sidewall

斜线　slant line/oblique line

卸货　cargo unloading

心理相容性　psychological compatibility

心理运动　psychomotor

锌　zinc

新陈代谢　metabolism

信号　signal

行李舱门　baggage door/luggage door

行李架　baggage shelf

行走道　walkway

型材　sections

型钢　steel shapes

性能　performance

性质　nature

修复性维修　corrective maintenance

修整　trimming/dressing

嗅觉　olfactory sensation

虚拟现实　virtual reality

虚线　dashed line/broken line/dotted line

需求确认计划　requirement validation plan

许用应力　allowable stress/permissible stress

旋转连杆　swivel link/rod

眩光　glare

巡航　cruise

巡视检查　walk-around check

迅速减压试验　quick decompression test

Y y

压溃纸管芯　crushed core

压敏胶带　pressure sensitive adhesive tape

压缩成形　compression moulding

压缩强度　compressive strength

亚声速飞机　subsonic aircraft

烟灰缸座　ashtray holder

烟雾探测器/放大器　smoke detector/amplifier

烟雾探测装置　smoke detection unit

严重损坏　significant damage

研究员级高级工程师　professor-level senior engineer

研磨料　abrasive

研制　development

研制方案　development scheme

研制规划　development programme

研制计划　development plan

研制网络图　development network drawing

验收试验　acceptance test

验收试验程序　acceptance test procedures

扬声器板插销　speaker panel latch
阳极化　anodizing
仰视图　upward view
氧化　oxidation
样板　template
样机图　phototype aircraft drawing
液压系统　hydraulic system
一般检查　general inspection
一般目视检查　general visual check
一字槽螺钉　minus screw
衣橱　wardrobe
医学监督　medical monitoring
仪表板　instrument panel
乙醇/酒精　alcohol/ethanol
乙烯　ethylene
意志　will
银色　silver
印刷品袋　literature pocket
应变　strain
应变集中　strain concentration
应急撤离地面标识灯　marking light
应急撤离指示灯　emergency evacuation light
应急出口装饰板　emergency exit furnishing
应急起动手柄　emergency trigger lever
应急着陆　emergency landing
应力/应激　stress
应力集中　stress concentration

英寸[①] inch

荧光灯 fluorescent lamps

硬度 hardness

油漆工艺规程 painting process procedure

游动螺母 floating nut

有机玻璃 acrylic plastic sheet

有机塑料 organic plastics

有色金属材料 non-ferrous material

有色金属合金 non-ferrous alloy

有效期 valid period

右视图 right view drawing

右仪表板 right instrument panel

预定维护/计划性维修 scheduled maintenance

预定维修间隔 scheduled maintenance check interval

预防措施 preventive action

预防性维修 preventive maintenance

预固化 precure

预警机 early warning aircraft

预浸处理 prepreg

阈值 threshold

原理图 schematic diagram

原位检查 on-aircraft check

圆头螺钉 round head screw/button head screw/dome head screw

圆头螺栓 round head bolt/cheese head bolt/button head bolt

圆头铆钉 round head rivet/cup-head rivet/globe head rivet/snap-head rivet

圆柱体 circular cylinder/cylinder

---

① 英美制长度单位，1 inch=25.4 mm。

圆柱头螺栓　cylindrical bolt/fillister head bolt

圆柱销　parallel pin/cylindrical pin

圆柱形中段机身　cylindrical midway through the fuselage

圆锥体　cone

远程飞机　long-distance aircraft

约等于　approximately equal to

月检查　monthly check

阅读灯　reading light

运输机　transport aircraft

运行　operating

Z z

杂项　miscellaneous

杂志盒/架　magazine rack

灾难级　catastrophic

载荷系数　load factor

早期固化　premature cure

噪声　noise

增强体　reinforcement/reinforce

增压舱载荷　pressurized compartment loads

增压机身　pressurized fuselage

粘接/胶接/工艺规程　bonding process procedure

粘接　bonding

胀接　expending joining

照度　illuminance

照明　illuminate

罩的耳朵型突出　lug of the cover

遮挡插条　cuttings of shield

遮光板　sunvisor

遮光罩 glareshield/lightshield
赭色 sienna
真空-压注法 vacuum-pressure injection
振荡 oscillation
振动 vibration
振动和冲击 vibration and shock
正方形 square
正火 normalizing
正驾驶员/机长 pilot/captain/first-pilot
支撑 jack
支承连杆 support link/rod
支线机 regional aircraft/commuter
知觉 perception
织物地板覆盖物 textile floor covering
直角三角形 right-angle triangle
直径 diameter
直升机 helicopter/rotorcraft
直线 straight line
止动块 stop/stopper
止裂孔 crack arrest hole
纸蜂窝材料(商标名) NOMEX
制造厂 manufacturer
制造方法 fabrication methods
质量保证 quality assurance
质量保证计划 quality assurance plan
智力 intelligence
置入 engage
中部 central section
中层玻璃 middle pane

中程飞机　medium-range aircraft

中短程飞机　short/medium aircraft

中机身/机身中段　mid fuselage/center fuselage

中间　middle

中间隔层　interlayer

中门半槽边　middle door reveal

中心　centre

中心线　center line

中央仪表板　center instrument panel

中央翼盒　wing center box

重量　weight

重量符合性报告　weight compliance report

重量控制大纲　weight control plan

重量限制　weight limitation

重量状态报告　weight status reports

重心　center of gravity

重要级　major

周(边)长　perimeter

轴　shaft

轴承　bearing

主舱地板　main deck

主货舱　main deck cargo compartment/master cargo compartment

主客舱　passenger compartment

主连杆　main link/master rod

主要几何尺寸　principal geometric dimensions

主要结构　primary structure

主要特征　leading particulars

主应力　principal stress

主坐标　principal coordinates

助理工程师　assistant engineer
注射成形　injection moulding
注意　attention
贮存期　shelf life/storage life
柱　column
铸件图　casting drawing
铸铁　foundry iron
铸造工艺　casting technology
专家　specialist/expert
转接接头　transfer joint
转台成形　rotary moulding
装货　cargo loading
装配/分解工艺规程　process procedure for assembly and disassembly
装配工艺　assembly technology
装配图　assembly drawing
装饰板　decorative panel
装饰膜　decorative Foil
着陆　landing
着陆后检查　after-landing check
姿态　posture
紫　purple/violet
自熄　self-extinguishing
自由成形　free forming
综合　combination
综合试验程序　integration test procedures
综合试验项目、大纲　integration testing items and plan
总工程师　chief engineer
总经理　president
总设计师　chief designer

总体布置图　overall arrangement/layout drawing

总体破坏　catastrophic failure

总图　general drawing

总装配　final assembly

纵剖线　buttock line

阻潮层　moisture barrier foil

阻尼　damping

阻尼器　damper/absorber

组装　set up

钻孔/削　drilling

最大/最小　maximum/minimum

左视图　left view drawing

左仪表板　left instrument panel

作动筒/器　cylinder/actuator

作用剂　agent

座舱隔框　cabin bulkhead

座排号　seat row number

座位号　seat numbers

座椅　seats

座椅地轨盖板　seat track cover

座椅地轨连接接头　seat floor track attachment fitting

生活设施

导 语

飞机机内生活设施指为机上人员提供必要的工作、生活服务保障的设施设备，以保障机上人员的舒适性、便利性和安全性。通常，机内生活设施包括座椅、厨房、卫生间、卧铺等设备。

飞机绝热层的主要功能是降低通过机体结构进入座舱内部的热量和噪声量。绝热层敷设在内饰板和蒙皮之间的空间区域。

一块典型的绝热层由包覆层包覆芯材，周圈经缝纫或热压密封制成。芯材通常为玻璃纤维或泡沫。

适航条款要求绝热层材料应满足阻燃要求，机体构造水平面以下敷设的绝热层还应满足抗火焰烧穿的要求。

通常，正驾驶员（机长）座椅和副驾驶员座椅具有座椅调节结构，能够调节座椅的前后位置。控制元件布置于座椅靠近舱内的一侧。座椅上部的调节机构能够控制座高、椅面角度、靠背角度、扶手高度、腰靠和头靠位置的调节。

座椅底部通过滑块导向组件安装于座椅导轨上。座椅通过滑块导向组件可以方便地调节前后位置。通常位置调节后，采用一个弹簧插销锁将座椅固定至导轨合适的位置上。

观察员座椅为观察员提供乘坐位置。通常观察员座椅包括：座椅靠背、安全带、座椅坐垫、金属骨架等。金属骨架用于支撑座椅坐垫。座椅坐垫由具有浮力的材料制成，经批准可用作水上迫降时的漂浮救生设备。观察员座椅设置有一套安全带。观察员座椅不使用时，座椅坐垫可折叠收起；使用时，将座椅坐垫放下供观察员乘坐。

乘客座椅通过座椅导轨固定在客舱地板上。每个乘客座椅设置一套安全腰带。座椅靠背角度可调节。大部分乘客座椅设计有可折叠式小餐桌。乘客座椅通常为双连座或三连座形式。乘客座椅坐垫可作为水上迫降时的漂浮救生设备，座椅下方放置有救生衣。

飞机厨房为机上乘员提供准备食品和饮料的设施设备。厨房由厨房结构、厨房电气和厨房插件组成。厨房结构包括厨房框架、抽拉式工作台面、储物柜、废物箱、扶手、支座和安装支架、机械接口、活动件约束、水龙头、水槽和滤网等附件。厨房电气包括电气控制电路、厨房控制面板、过流保护装置、电连接器和电气接口。

不同的厨房构型会设置不同的厨房插件。典型厨房插件包括:厨房冷冻器、烤箱、冰箱、咖啡壶、水槽、储存箱、垃圾箱、服务手推车等。

厨房的安装通过顶部安装件与飞机结构连接,通过底部支座与飞机结构固定。

厨房控制面板用于控制厨房电源供应和电气设备的工作状态。

盥洗室应满足所有可能类型乘员如厕及洗漱功能的需求。盥洗室包含盥洗室结构和盥洗用具两部分。盥洗室结构包括隔墙、天花板、地板、门、马桶装饰罩、衣帽钩、扶手、烟灰缸、安装连接件等。盥洗用具包括洗脸盆、梳妆台、洗涤水龙头、梳妆镜、擦手纸盒、手纸盒、马桶垫纸盒、盥洗室婴儿托架、皂液器盒、垃圾箱、乘务员呼叫按钮、镜前灯灯罩、微动开关等。

The aircraft interior equipment provides necessary working and life service equipment to passengers and crews, in order to supply safety、comfort and convenience to all occupants. Generally, interior equipment includes seat, galley, lavatory, sleeping berth and so on.

Aircraft Insulation blanket reduces sound and heat transfer through the fuselage walls and attached between the interior linings and fuselage skin. A typical blanket consists of a layer of core material, normally fiberglass or foam. an inboard cover, and an outboard cover. The two covers are stitched or cemented together at the blanket edges.

The material of the insulation blanket shall meet the fire-resistant and insulation in the lower half of the fuselage shall meet the burn through requirement according to the airworthiness regulation.

Generally, the captain and first officer seat have controls and mechanisms for seat fore/aft position. The controls are on the inboard side of each seat. The upper seat has controls and mechanisms for seat height, thigh pad position, seat recline, armrest height, back cushion (lumbar support) position and headrest position adjustments.

Bogie units hold the base to the aircraft seat tracks. Rollers in each bogie unit make adjustment of the seat position easy. Normally, a spring-loaded track lock mechanism sets fore and aft movement on the seat tracks.

The observer seats supply crew stations for extra crew members. Generally, the observer seat has these parts: seat back, safety harness, seat bottom cushion, metal frame, and so on. The metal frame holds the seat bottom cushion. The seat bottom cushion is a buoyant material and is an approved flotation device. The seat has a set of safety harness. The seat folds to the flight compartment wall when not in use. To use the seat, lower the seat into position.

The passenger seats attach to the seat tracks in the floor. Each seat has a lap belt. The back of each seat reclines.

Most seats have trays which fold down.

The seats are two-or three-passenger assemblies. the seat cushions can used as flotation devices and life vests stowed in the space under the seats.

Galleys provide food and beverage preparation facilities. Galley is composed by galley structure, galley electrical subsystem and galley inserts. The galley structure mainly includes composite panel, pull-out worktable, storage compartment, waste bin compartment, handhold, attachment bracket and joint, mechanical interface definition, constrains and miscellaneous (faucet, sink and filter). The gal-

ley electrical system includes galley electrical control circuit, electrical control panel, overcurrent protection device, connector and electrical interface.

Galleys may have different inserts. The typical galley inserts include: chiller, oven, refrigerator, coffee maker, sink, storage, waste container, serving carts and so on.

Upper fittings attach the top of the galley to the airplane structure. Floor fittings attach the bottom of the galley to the airplane structure.

Generally, the electrical control panel is used to control the galley power supply and the electrical device working state.

The lavatory shall meet the toilet and the washing requirements for passengers. The lavatory includes the structure frame and the washing appliance. The lavatory structure includes: clapboard, ceiling, floor, door, toilet decorate cover, coat hook, handhold, ashtray, installment fittings. The lavatory appliance includes: washbowl, counter table, faucet, mirror, tissue box, toilet paper box, toilet seat paper box, baby nursing table, soap dispenser, waste bin, attendant call switch button, mirror light cover, micro-switch and so on.

缩略语

A a

AD Accidental Damage 偶然损伤

ALI Airworthiness Limitation Items 适航限制项目

AOB Any Other Business 任何其他事项

AS Aerospace Standard 航空标准

ATD Anthropomorphic Test Devices 仿真试验假人

C c

CAMI Civil Aeromedical Institute 民用航空医学研究所

CDR Critical Design Review 详细设计评审

D d

DOC Direct Operating Cost 直接使用成本

DR Dispatch Rate 派遣率

DTA Damage Tolerance Analysis 损伤容限分析

E e

EASA European Aviation Safety Authority 欧洲航空安全当局

EMI Electromagnetic Interference 电磁干扰

EPS Electrical Power System 电源系统

F f

FAA Federal Aviation Administration 联邦航空管理局

FAR Federal Aviation Regulations 联邦航空管理条例

FC Flight Cycle 飞行循环

FDAL　Functional Development Assurance Level　功能研制保障等级

FH　Flight Hours　飞行小时

FHA　Function Hazard Analysis　功能危害分析

H h

HIRF　High Intensity Radiated Fields　高强辐射场

I i

ITCM　Initial Technical Coordination Meeting　技术协调启动会

L l

LLI　Life Limitation Items　时寿项目

LORA　Level of Repair Analysis　修理级别分析

LRM　Line Replaceable Module　航线可更换模块

P p

PPH　Policy,Procedure Handbook　政策,程序手册

专业术语

A a

安装点 attachment point

安装检查 installation-check

安装人员 installation applicant

B b

把手 handle

保证/保障 assure

报纸杂志袋 literature pocket

杯子投放器 cup dispenser

备忘录 memorandum

背离/偏离 deviation

便于佩戴及解脱 be easy to don and remove

变形 deformation

标准 standard

标准型五点式约束系统 standard five-point restraint harness

标准装载单元(装置) standard carry on units

表面包覆膜 cover foil/covering film

冰箱(柜) refrigerator

玻璃纤维表面覆层 fiberglass cover plate

不紧的/松弛的 slack

不利影响 adverse effect

不情愿去做某事 be reluctant to do sth.

不锈钢踢脚线 stainless-steel kick strip

不一致/不协调　inconsistency

手柄　handgrip

C c

擦手纸存取箱　towel disposal container

餐具架　dish rack

参与/参加　participate

侧杆扶手　side stick armrest

侧荷-位移　lateral load-displacement

测量　measurement

侧门　side door

测试程序　test procedure

测试和评估标准　test and evaluation criteria

测试件　test article

侧向座椅的安全带　side-facing harness

厕纸卷筒架　toilet paper roll holders

插头　plug

插头座　connector

插销手柄　plunger handle

插座　receptacle

撤离　evacuation

成本-效益　cost-benefit

乘务员座椅　attendant seat

乘员保护　occupant protection

乘员环境危险　occupant environment hazards

乘员加速度环境　occupant acceleration environment

乘员身材体态　occupant morphology

乘员生还可能性　occupant survivability potential

乘员危险坠撞下的耐受极限　occupant exposure limits to crash hazards

乘员响应(克服)曲线　occupant response (ride-down) curve

乘员约束系统强度要求　occupant restraint harness strength requirements

乘坐空间的防护　preservation of occupied space

齿条齿轮组合件　rack and pinion assembly

冲程　stroke capacity

冲击保护　impact protection

冲击力的方向　orientation-to-impact forces

冲击能量管理　impact energy management

充气式身体和头部约束系统　inflatable body and head restraint system

抽水马桶　flush toilet

厨房　galley

厨房/卫生间构型定义　monuments configuration definition

厨房地板防溅板　galley floor splash guard

厨房地板覆盖层 galley floor cover

厨房隔框单元　galley compartment

厨房接地装置　galley grounding

厨房冷却空气　galley chilled air

厨房门　galley door

厨房配套设施　galley auxiliary facilities

厨房升降机驱动系统　galley lift drive system

厨房升降机系统　galley lift system

厨房照明　galley lighting

厨房制冷系统　galley cooling system

储藏舱保温器　bunker warmer

储藏柜　storage

触头　contact

传动链　chains

垂向速度改变量　change in vertical velocity

垂直调节作动筒　vertical actuator

粗心导致的/疏忽造成的　inadvertent

D d

搭扣　buckle

带滚轮座椅　roller seat

带软垫的/舒适的靠背　upholstered backrest

带子/带状织物　webbing

单点释放系统　a single-point release system

单人座椅　single seat

单叶门　blade door

导轨凸缘　track flange

导致　contribute to

等距的　equidistant

底盘　floor pan

地脚螺栓　stud

第三机组座椅　third occupant seat

第五点安全带　fifth strap

典型飞行员重量　typical aviator weights

电动剃须刀电源　electric razor power supply

电动剃须刀电源插座　electric razor outlet

电动座椅　electrical controlling seat

电加温靠背　electric equipment backrest

电线接头　electrical connection

吊式座椅　hammock-type seat

动态冲击测试　dynamic impact test

动态冲击脉冲要求　dynamic crash pulse requirement

动态试验参数　dynamic test parameters

独立单元　autonomous element
断裂强度/破坏强度/抗断强度　breaking strength
对抗/阻碍/抵制/抵消/中和　counteract
带包络面特征的数模　envelope model
待定事项　pending item
导管支架　duct support
导轨　guide rail
导向装置　guide unit
地板排水装置　floor drain
电磁系统　electromagnetic system
电动升降机　electrically powered lift
电机/减速器组件　motor/gearbox assembly
电气接口　electrical interface
电气连接图　wiring diagram
电线　wire
电线线束　electrical harness
电源控制　power control
定量要求　quantitative requirement
断路器　circuit braker
惰轮组件　idler sprocket assembly

F f

返回座位标识　return to seat sign
方法/手段　recourse
方向　orientation
防火　fire-resistant
防溅板　splash guard
防烟镜存放处　smoke goggle stowage
防止乘员被局部卡住　localized entrapment prevention

妨碍　hinder/impede

飞机寿命周期　aircraft life cycle

飞机坠撞防护范围　aircraft crash protection envelope

飞机坐标系与姿态方向　aircraft coordinate and attitude directions

飞行工程师座椅/机械师座椅　flight engineer's seat

废料车　waste cart

废料箱　waste bin/garbage container

废物箱　waste compartment

废物箱弹簧门　waste compartment spring loaded door

分阶段鉴定　grading verification

分配　allocation

分配/分派　assign

缝纫　stitching

扶手调节器　armrest adjuster

服务用品橱阁　service cabinet

服役的能量缓冲装置　operational energy absorbers

符合性报告　compliance report

符合性核对清单　compliance check list

腹带防护对比肩带-腹带防护　comparison of lap belt only versus lap belt - shoulder restraint protection

腹带固定点　lap belt anchorage

复合材料　composite material

复位弹簧　return spring

负载　load

G g

改装翻新　retrofit

干厨房单元　dry galley unit

甘特图　Gantt plot

刚性测试试样　rigid test block
刚性硬件　rigid hardware
高低可调座椅　vertically adjustable seat
高度调节　height adjustment
高分子塑料支架　polymer plastic stents
高熔点芳香族聚酰胺　nomex
搁脚处　footrest
隔热层　thermal insulation layer
隔热隔声　thermal and acoustical insulation
隔热隔声泡沫材料　insulation foam
隔热隔声毯/隔热隔声层　insulation blanket
隔声层　sound insulation layer
更衣室　dressing room
工程合作方　engineering counterpart
工程数据　engineering data
工作台　work station/work counter
功能接口　functional interface
功能模块　functional block
构型　configuration
构型管理计划　configuration management plan
构型控制程序　configuration control process
股/大腿　thigh
骨盆/盆腔　pelvis
固定橱柜　fixed cabinet
固定式座椅　fixed seat
故障隔离控制板　fault isolation panel
挂衣钩　coat hook
关键词列表　subject term (key word) listing
关键区域确认　identification of potential critical areas

贯彻/实行　implementation

惯性卷筒装置　inertia reel

盥洗池、洗脸池　washbasin

盥洗间　washroom

盥洗室发光指示牌　lavatory lighted-sign

盥洗室灭火器　lavatory fire-extinguisher

盥洗室烟雾探测器　lavatory smoke detection

盥洗室照明　lavatory lighting

规定试验和评估判据　defining test and evaluation criteria

规范、说明　specification

滚珠丝杠组合件　screw-ball nut assembly

过度设计　over-design

过度移动　excessive motion

过热继电器　overheat relay

过温传感器　overtemperature sensor

过载保护装置　overload protection device

H h

合同双方达成一致　mutual agreement

轰炸员座椅　bombardier seat

后向座椅的安全带　aft-facing harness

呼叫按钮　call button

花键轴　spline shaft

滑车系统　sled system

活动橱柜　removable cabinet

货舱绝热层　cargo compartment insulation

货物和辅助装备的系留固定　cargo and ancillary equipment retention

货物吸能约束带的前向载荷-位移　forward load-displacement for

energy absorbing cargo restraint

J j

机上娱乐系统　in-flight entertainment(ife)system

机体防护　airframe protective shell

机械调节式座椅　mechanical controlling seat

机械接口　mechanical interface

基座式盆腔　floor well

几何包络面　geometrical envelope

挤压型材　extrusion

脊柱方向加速度的持续时间和级别大小　duration and magnitude of spineward acceleration

技术维护规程　maintenance instruction

继电器　relay

加热水箱　water boiler

加速度损伤防护　acceleration injury protection

夹层板　sandwich panel

驾驶舱座椅设计和静态试验要求　cockpit seat design and static test requirements

肩带固定点　shoulder harness anchorage

肩带和腹带约束的乘员碰撞包线　occupant strike envelope with shoulder harness and lap belt

减速开关　deceleration switch

减震器　shock absorber

减震座椅　shock-absorption seat

简约损伤评级与身体两个部位的损伤类型　abbreviated injury scale (ais) and sample injury types for two body regions

交付物交付后支持　post-delivery support

角度调节　angular adjustment

接触器　contactor

接口定义　interfaces definition

接线盒　junction box

节点/期限　time frame

结束/终止　termination

解决(问题、争端、争论等)　resolve

解释/说明　interpretation

解锁控制装置　unlocking control

金属结构骨架　metal frame

仅有腹带约束的乘员碰撞包线　occupant strike envelope with lap belt only restraint

紧固件/夹持器　fastener

紧固螺栓　securing bolt

进气格栅　air inlet grill

经核准的/被认可的　approved

救生背心存放处　life vest stowage

救生衣检查窗　life vest check window

绝缘垫圈　insulation spacer

绝缘胶　insulation paste

绝缘漆　insulation paint

军用抗坠撞座椅系统　military crash seating system

军用座椅地板与舱壁变形图　floor and bulkhead misalignment diagrams for military seats

K k

咖啡制作具　coffee maker

开动(机械等)/驱使　actuate

开关　switch

开口项　open item

开口项确认　identification of open items

抗坠撞结构特点　crash resistant feature

抗坠撞座椅　crash resistant seating/crashworthy seat

烤箱　oven

靠背(倾度)可调座椅　reclining seat

靠背罩　backrest cover

可调节座椅　adjustable seat

可后仰靠背　reclining backrest

可利用的/可获得的　available

可能造成的后果　possible implication

可伸缩式立柱　telescopic column

可通过口盖　access flap

可卸式座椅　demountable seat

可旋转座椅　rotating seat

可折叠小桌板　folding table

空降人员座椅　troop seat

控制操作手柄　control lever

控制钢索　control cable

控制面板　control panel

快卸式安装方式　quick-release type fitting

快卸支座　quick release fitting

L l

垃圾斜槽/滑道　waste chute

垃圾压缩机　trash compactor

拉杆连接　tie rod attachment

冷柜　cold storage

冷却空气系统　chilled air system

冷却空气系统导管　chilled air system duct

冷却空气系统排水装置　chilled air system drain

冷却器　chiller/freezer/refrigerator

里程碑　milestone

利与弊　risks and benefits

联合分析和评估　joint analysis and evaluation

脸盆　washbasin

领航员座椅　navigator's seat

隆起　upheaval

漏水池　sink

旅客座椅　passenger seat/cabin seat

M　m

马桶　toilet

马桶便盆　toilet bowl

马桶冲洗按钮　toilet flush switch

马桶冲洗控制开关　toilet flush control switch

马桶冲洗手柄　toilet flush handle

马桶盖　toilet cover

马桶外罩　toilet shroud

马桶坐垫　toilet seat

马桶坐垫罩纸盒子　toilet seat cover dispenser

毛巾/食品加热器　burn warmer

铆钉/铆接　rivet

门(窗)帘盒/狭长木框　pelmet

门槛/门口　threshold

门锁　door latch

N　n

耐久性　durability

能量缓冲座椅　energy attenuating seat

黏合　bond

尼龙胶条　nylon tape

P　p

排号　row number

排气管子　air extraction hose

排水装置　drain unit

喷嘴式饮水龙头　drinking fountain spigot

碰撞损伤防护　contact injury protection

批准　approval

批准/确认　validation

疲劳　fatigue

评估　assessment

Q　q

其他可选择的建议方案　alternative proposal

启动/发起　initiate

气缸　gas cylinder

气囊系统　air bag system

签字　sign off

前后调节器　fore and after adjuster

前限位开关　forward limit switch

前向座椅的安全带　forward-facing harness

潜在的碰撞危害　potential strike hazards

切开　cutout

倾斜锁组件　recline lock assembly

取证要求　qualification requirement

取证支持计划　certification support plan

全尺寸坠落试验　full-scale drop test

确认/认可　confirm

R r

热杯　hot cup

热绝缘系数　heat insulation coefficient

热熔胶结　hot welding

热水壶　hot jug

热水器　water heater

人机功效设计　ergonomic design

人体测量学　anthropometry

人体对输入加速度的动力响应　human kinematic response to input accelerations

人体对突发加速度的耐限　human tolerance to abrupt accelerations

人体耐限　human exposure limits

人体损伤耐限　human injury tolerance

熔断器　fuse

容纳/适应　accommodate

容器　container

入口　entryway

S s

三连座座椅　triple seat

三位置控制开关　three-position switch

上/下服务组件　upper/lower service module

上部/下部滚轮组件　upper/lower roller assembly

上舱厨房　upper deck galley

上降机箱下降指示灯　cage “down” light

上锁销　locking pin

设备架　equipment rack
设备清单　list of equipments/appliances
设计不足　under-design
设计冻结　design freeze
设计审核清单　design checklist
设计输出　design output data
设计输入　design input data
设计准则指南　design criteria guidance
升降调节器(垂直)　vertical adjuster
升降机井道　lift well
升降机门　lift door/cage door
升降机门闩　lift door latch
升降机限位开关　lift limit switch
升降机箱　lift car/lift cage
升降机箱连接组件　cage attachment assembly
升降机箱内部控制开关　inside cage control switch
升降机箱上升指示灯　cage up light
升降机应急出口　lift escape hatch
升降限位开关　up/down limit switch
生存和救生设备　survival and rescue equipment
湿厨房单元　wet galley unit
实验室试验方法　laboratory test procedure
食品/饮料车　food/beverage cart/trolley/carriers
食品/饮料车存放处　cart stowage
食品车固定器　cart restraint
食物和废物管理系统　sustenance and waste management
食物贮存柜　foods storage
使不动/使固定　immobilize
使用极限静载荷　statically applied ultimate load

士兵和炮手重量　troop and gunner weights
士兵座椅约束带的构型　troop seat restraint configurations
试验计划　test plan
试验假人的校验　test dummy calibration
适坠性对比　crashworthiness comparison
适坠性权衡选项　crashworthiness trade options
收藏式座椅　storable seat
手册　handbook
手驱动软轴　manual drive flexible shaft
手驱动手柄　manual drive handle
手纸箱　toilet paper box
首件　first article
首件交付物　the first deliverable
梳妆柜　dressing cabinet
梳妆台　dressing table
数值刻度　numerical scale
双连座椅　twin seats
双模动态响应　dual mode dynamic response
双人座椅　double seat
水池　basin
水冷却器　water cooler
水龙头　water faucet
水平调节器　horizontal adjuster
水平调节作动筒　horizontal actuator
四连座座椅　quadruplet seat
塑性变形 plastically deform
随机服务员座椅　cabin attendant's seat
损伤机理　injury mechanisms

T t

套筒　sleeve

提纲　outline

提高/增加/加强　enhance

提起插销手柄　raising plunger handle

提升/改进　upgrade

调节控制手柄　adjustment control handle

调整手柄　adjustment handle

通过铰接连接到……　be hinged to

通信员座椅　communicator's seat

通气孔　ventilation hole

头部方向加速度的持续时间与大小　duration and magnitude of headward acceleration

头靠　headrest

头靠垫　headrest cushion, headrest pad

凸出物　projection

推车升降机箱　cart lift enclosure

推荐的军用损伤评估试验方法　recommended military injury assessment test methods

推荐的军用腰椎耐受水平　recommended military lumbar tolerance levels

W w

往复运动行程　reciprocating motion distance

外壳/外罩　housing

外廓包边型材　edge profile

完善/改进　refine

完善改进建议/提议　improvement/modification proposal

微动开关　microswitch
维修任务分析　maintenance task analysis
维修性　maintainability
维修性分析报告　maintainability analysis report
维修止动块　maintenance stop
卫生间构件　lavatory structure
卫生间镜子　lavatory mirror
卫生间门　lavatory door
卫生间组件　lavatory module
卫生间座椅　lavatory seat
稳定滚轮组件　stabilization roller assembly
问题/议题　issue
五点式约束带构型　five-point restraint harness configuration
误操作释放　inadvertent release

X x

吸能装置　energy absorption device
吸能座椅系统　energy absorbing seat system
系留系统的强度　retention system strength
系统功能单元　system functional element
系统架构　system architecture
系统描述　system description
限制，约束　restrain
相对频度　relative frequency
相对速度　relative velocity
相关的　relevant
相关风险确认　identification of associated risks
项目产品支持计划　project product support plan
项目管理计划　project management plan

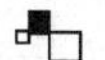

项目鉴定取证计划　qualification program plan

项目进展　project progress

项目主进度计划　project master schedule

橡胶条　rub-strip

销子/钉子　pin

小柜　cabinet

协调会议议程　coordination meeting agenda

斜坡设计方法　ramp design

信息交换　information exchange

性能要求　performance requirement

悬殊/迥异/差异/不符　discrepancy

旋钮　knob

旋转调节　rotation adjustment

Y y

烟灰缸　ashtray

严重的　severe

验证、核实　verify

扬声器　speaker

腰部调节　waist adjustment

腰部调节器　lumbar adjuster

腰带　abdominal belt

腰靠　lumbar rest

腰椎载荷损伤判据　spinal lumbar load injury criterion

液体肥皂配置器　liquid soap dispenser

一致性检查　conformity inspection

衣帽钩　coat hanger

移动……的位置　displace

椅面组合件　seating assemble

椅盘　seat pan

椅盆　seat bucket

易卸座椅　easily removable seat

引用文件　applicable documents

饮食服务中心　food and beverage service center

饮水处　drinking station

应急释放要求　emergency release requirement

应急锁紧装置　emergency locking retractor

应急照明　emergency lighting

婴儿托架　baby nursing table

硬点连接　hardpoint attachment

用电设备　utilization equipments

与……互搭、与……重叠　overlap

与……相符合　compliant with

与……一起　along with

预期可生还坠撞冲击条件　anticipated survivable crash impact conditions

遇到　encounter

约束类型及几何外形　restraint type and geometry

Z z

杂物袋　garbage bag

在……下方　beneath

皂液器　soap dispenser

粘接胶带　adhesive tape

展开　deploy

折叠门　folding door

折叠式座椅　folding seat/collapsible seat

振动开关　toggle switch

整理综合　consolidate

整体的乘员防护系统　integrated occupant protection system

正常操作载荷　normal operational load

证明/鉴定　certification

支架　bracket

织物加强型　scrim reinforced

执行　carry out

质量保证计划　quality assurance plan

致命的/灾难的　fatal

种类/类别　category

重量估算　weight estimate

重量控制计划　weight control plan

逐点　point-by-point

主动轮组件　drive sprocket assembly

主机限制　host restriction

主客舱厨房　main deck galley

主要因素　the dominant factor

助力手柄　assist handle

转椅　swiveling seat

坠机后的损伤防护　post-crash injury protection.

坠撞冲击设计条件　crash impact design condition

坠撞救生　crash survivability

坠撞可生存性通用设计要素　general crash survivability design factors

资料更新　supersession data

字母顺序刻度　alphabetical scale

综合分析　collective analysis

综合限制条件　integration constraints

总结报告　summary report

总重和任务 gross weight，and mission

阻燃、耐热的包覆层 heat resistant cover

钻孔支座 drilled fitting

最大可接受的损伤水平 maximum acceptable injury level

最低性能要求标准 minimum performance standard

最佳的支撑 optimum support

最起码文件要求 minimum documentation requirements

最少文件/使用说明书要求 minimum documentation requirement

最小弹性回弹 minimum elastic rebound

坐垫 seat cushion

坐垫材料推荐的应力-应变 recommended stress-strain corridors for cushion materials

坐垫泡沫 cushion foam

坐垫压力分布 cushion pressure distribution

坐垫罩 seat cushion cover

座位配置 seat configuration

座椅安全带 seat safety belt/shoulder harness

座椅弹射器 seat catapult

座椅导轨 seat track/seat rack

座椅底座 seat base/seat pedestal

座椅底座角护 base corner shroud

座椅电动机 seat gear motor

座椅电源供应 in-seat power supply

座椅调节 seat control/adjust

座椅调节器 seat adjuster

座椅断路器板 seat circuit breaker panel

座椅扶手 armrest

座椅供电电缆 seat supply wire

座椅供应商 seat supplier

座椅固定插销(棒)　seat plunger
座椅轨道　seat track
座椅滚轮衬套　roller bushing
座椅滚轮垫圈　roller washer
座椅后倾调节手柄　seat recline control lever
座椅夹紧器　seat gripper
座椅结构　seat structure
座椅结构的完整性及乘员的固定　seat structural integrity and occupant retention
座椅救生衣存放处　seat life vest compartment
座椅靠背　seatback seat, backrest
座椅控制电路　seat control circuit
座椅排列　seating arrangement
座椅前/后导轨　fore/after track
座椅前后调节开关　seat fore and after control switch
座椅升降调节开关　seat vertical control switch
座椅外形　seat shape
座椅(外)罩　seat outer cover
座椅向前载荷变形的要求　seat forward load-deflection requirements
座椅转盘　seat rotating star
座椅坠撞载荷系数　seat crash load factors
座椅纵向-横向调节手柄　seat longitudinal-lateral adjust lever

环控救生系统试验

导 语

一般来说，建立飞机环境控制和救生系统(简称环控救生系统)实验室是为了模拟工作环境以及测试飞机环控系统和救生系统的功能和性能。它负责一架飞机生命周期内所有子系统的测试、附件测试和系统联合测试，包括了各种型号飞机的环境控制系统测试、救生系统测试、环境保护系统测试和氧气系统测试。

典型的飞机环控救生系统实验室应该由引气系统、加热冷却系统、高海拔环境仿真系统、测控系统和其他必要的子系统组成。引气系统和加热冷却系统可以模拟发动机的引气和冲压空气。高海拔仿真系统提供各种参数的仿真，包括高度、温度、湿度等。测控系统由传感器系统、数据采集系统和监控系统组成，可以完成测试数据的测量、过程控制和设备监控工作。

Generally, the Aircraft Environment Control and Lifesaving System laboratory is established to simulate the working environments and to test the function and performance of the Aircraft Environment Control and Lifesaving System. It shall be responsible for all of the subsystems test, accessory test and system joint test in the whole lifecycle of a aircraft, involved Environment Control System test, Lifesaving System test, Environmental Protection System test and Oxygen System test of types of aircraft.

A typical Aircraft Environment Control and Lifesaving System laboratory shall consist of gas resource system, heating and cooling system, high-altitude environment simulation system, measurement and control system, and other necessary subsystem. Gas resource system and eating and cooling system shall simulate the aircraft engine bleed air and ram air. High-altitude environment simulation system provides simulation of various parameters including height, temperature, humidity, etc. Measurement and control system, which is composed of sensor system, data acquisition system and monitoring system, shall complete test data measurement, process control and device monitoring.

专业术语

A a

安全灯　safety lighting fitting

安全泄气阀/安全阀　safety valve, SV

安全泄压阀　safety relief valve

安全泄液阀/卸荷阀　relief valve

安装　erection

安装高度　mounting height

鞍座　saddle

按钮　push-button

暗装单极板钮开关　flush-mounted single-pole toggle switch

暗装双极插座　flush-mounted 2-pole receptacle

凹面　female face,FMF

凹凸面　male-female(Seal contact)face, MFM

奥氏体不锈钢　austenitic stainless steel

奥氏体不锈钢管　austenitic stainless steel pipe

B b

白炽灯具　incandescent lamp (bulb)

白金属　white metal

百分表　centigrade gauge

板表面　plat face

板翅式换热器　plate-fin heat exchanger

板卡　card

板框式压滤机　plate and frame filter press

板片换热器　plate type heat exchanger

板式平焊法兰　welding plate flange

板式起重机　bridge crane

板式塔　plate tower

版次　issue

半导体二极管　semiconductor diode

半径样板　radius template

半永久性真空封接　semi-permanent seal

半圆锉　half-round file

伴热　heat tracing

包装　packaging

保安电源　emergency power supply (electric source)

保安电源送电　supply of emergency electric source

保持触点　holding contact

保护掉闸　protective trip

保护管　protective tube

保护接地　protective earthing

保护设备　protective device

保护元件　protective element

保护装置　protective device

保冷　cold instruction

保温　insulation

保温箱/保护箱　heating box/protection box

爆破板　rupture disk

爆炸性环境　explosive atmosphere

备件　spare part

备忘录　memorandum

备用回路　spare circuit

备用母线　spare bus
备注　remarks
背部接线图　back wiring (diagram)
背视图　back view
背压　back pressure
被调量　controlled variable
本质安全电路　intrinsically-safe circuit
本质安全型设备　intrinsically-safe apparatus
泵房　pump house/pump room
比降-面积法　slope-area method
比例　scale
比例积算器　proportional accumulator/proportional totalizer
比热比　ratio of specific heat capacities
比容　specific volume
比重　specific gravity
闭环控制　closed loop control
壁灯　wall light
壁厚　wall thickness
壁厚系列号/管子表号　schedule number/Sch. NO.
壁温　wall temperature
篦子板　grate/grating
避雷带　strap type lightning protector
避雷器　lightning arrester/surge discharger
避雷网　network of lightning protector
避雷针　lightning rod
避雷针尖　tip of lightning rod
避雷针拉铁　brace for lightning rod
避雷针支架　lightning rod support

避雷装置　lightning protector

边界内　inside battery limit

边界外　outside battery limit

边屏　side board

扁钢　flat bar

扁平锉　flat file

变电所　substation

变电所紧急停车　emergency shutdown at substation

变频器　variable-frequency drive

变送器　transmitter

变形　deformation

变压器　transformer

变压器室　transformer room

标定量程　calibrated span

标高/高程/立面图　elevation

标准　standard,STD

标准表法　master meter method

标准大气压力　standard atmosphere

标准件　standard part

标准弯管　normal bend

表面处理　surface treatment

表面粗糙度　degree of finish

表面闪络　surface flash-over

表面涂覆　surface coating

丙烯腈-丁二烯-苯乙烯橡胶　acrylonitrile-butadiene-styrene rubber

波长表　wave-length meter

波纹管　bellows/corrugated tube

波纹管密封阀　bellow sealed valve

波纹膨胀节　bellow expansion joint

玻璃布　glass(fiber) cloth

玻璃分级过渡封接　glass graded seal

玻璃管　glass tube

玻璃管式压力计　glass tube pressure gauge

玻璃棉　glass wool

玻璃纤维增强聚酯　glass fiber reinforced polyester

薄孔板　thin orifice plate

补偿导线　extension wire

不饱和聚酯　unsaturated polyester

不合格品　defective unit；non conforming article

不间断电源　uninterrupted power supply，UPS

不连接的跨越导线　crossing wires not in contact with each other

不确定度　uncertainty

不锈钢　stainless steel

不锈钢管　stainless steel valve

不许出现袋形　not allow pocket

布袋收尘器　cloth envelop collection

部门　department

C　c

采样　sampling

采样时间　sampling time

采样速率　sampling rate

采样信号　sampled signal

参比信号　reference signal

参考/基准　reference

参考图　reference drawing

残差　residual error

残留电压　residual voltage

残余气体谱　residual gas spectrum

残余压力　residual pressure

操作方式选择开关　selecting switch for types of operation

操作小母线　operating miniature bus

槽钢　channel

槽面　groove face,GF

草图　sketch

测功器　dynamometer

测井　gauge well tiling

测量　measurement

测量电极　meter electrodes

测量范围　working range

测量管道　measuring piping

测量仪器的禁用　rejection of a measuring instrument

测量值　measured value

测温点　temperature measuring point

层流流量计　laminar flowmeter

插板阀　gate valve

插接器　plug-in connector

插入式　insert-type

插头　plug

插座　socket

插座干线　receptacle main line

插座数　number of receptacles

差动保护回路　differential protective circuit

差动继电器　differential relay

差流继电器　differential current relay

差压　differential pressure

差压变送器　differential pressure transmitter

差压流量计　differential pressure flowmeter

柴油发电机　diesel generator

缠绕式垫片　spiral wound gasket

长半径弯头　long radius elbow

长度　length

长径喷嘴　long radius nozzle

长期载流量　continuous carrying capacity

长时间额定值　longtime rating

厂区道路照明灯　street lighting in plant area

厂商协调会　vender co-ordinate meeting

场地　yard

超高真空阀　ultra-high vacuum valve

超临界压力　supercritical pressure

超声波换能器　ultrasonic transducer

超声波流量计　ultrasonic flowmeter

超声波式料位计　ultrasonic level meter

超声波探伤　ultrasonic test

车道　drive way

车螺纹　threading/thread turning

车削　turning

沉头螺栓　countersunk(head) screw

衬里管　lined pipe

成品版设备布置图　production plot plan

成品设计阶段　production design phase

成套设备　package unit

承包　contractor

承插的　belled

承插焊的　socket welded

承插焊法兰　socket welding flange

承插焊管帽　socket welded cap

承插焊管座　socket welded let

承插焊连接　socket welding flange

承插连接/套筒连接　bell and spigot joint

澄清器　gravity settler

迟滞　hysteresis

齿轮泵　gear pump

充电屏　charging panel

充分发展的速度分布　fully developed velocity distribution

充气阀　charge valve

冲击电压　impulse voltage

冲击继电器　impact relay

冲量式质量流量计　impulsive mass flowmeter

冲洗阀　flush valve

冲压件　stamping

重复接地　multiple earthing

重复性　repeatability

重合闸继电器　reclosing relay

抽气封口接头　pumping stem

抽气量　throughput of a pumping unit

抽气时间　pump-down time

抽速　volume flow rate of a pumping unit

抽样检定　verification by sampling

出口　outlet/discharge

出口温度　outlet temperature

出口压力　outlet pressure

出线端子示意图　schematic diagram of terminal outgoing lines

出线套　outgoing line sleeve

初步阶段　preliminary stage

除尘器/吸尘器　duster/dust arrester

触电　electric shock

穿板接头　bulkhead union

穿线盒　pull box

传播时间式超声波流量计　time of flight ultrasonic flowmeter

串轴保护/轴向位移保护　rotor position protection

垂线　vertical

垂线流速分布曲线　vertical velocity curve

垂直/正交　perpendicular

垂直安装　vertical installation

垂直度　perpendicularity

醇酸瓷漆　alkyd enamel

瓷漆　enamel varnish

磁场　magnetic field

磁粉探伤　magnetic particle test

磁力泵　magnetic transmission pump/magnetic coupled pump

磁力起动器　magnetic starter

磁力起动器组　magnetic starter group

磁性流量计　magnetic flowmeter

粗抽管路　roughing line

粗抽时间　roughing time

粗加工　roughing

粗滤器　strainer

粗牙螺纹　coarse thread

粗制的　coarse

淬火　quenching

萃取器　extractor

存放柜　store chest

𝒟　𝒹

搭焊 lap welding

大电流发生器　strong current generator

大小头　reducer

带接换装置的蓄电池组　accumulator battery with tap-changers

带螺纹　thread end

带式(皮带)运输机　ribbon conveyer; belt conveyer

带式过滤机　belt filter

带丝堵三通　plug tee

带状卡　strap clamp

待定　hold

袋形管　bag-shape pipe

单回路的　single loop

单片机　microcontrollers

单声道斜束式超声波流量计　single-path diagonal-beam flowmeter

单速手摇绕线机　single speed hand winding machine

单头扳手　single-end spanner

单头活接头　half union

单线回路　single-wire circuit

单线图　one line diagram

单相电度表　single-phase kilowatt-hour meter

单相接地　single phase earthing

单相三孔明插座　surface-mounted single phase 3-pole receptacle

单相自耦变压器　single-phase auto-transformer

挡板　damper

挡板阀　baffle valve

刀开关　knife switch

导电度　conductivity

导热系数　thermal conductivity factor

导线钳压器　wire jointing press-clamp

导线引上　conductors turning up

导线引下　conductors turning down

导线由上引来　conductors turning from above

导线由下引来　conductors turning from below

导向板/夹板　cleat

导向架　guide

导压管　pressure pipe

倒角　chamfering

倒圆角　rounding/filleting

德国工业标准　Deutsche Industrie Norm,DIN

灯数　number of lamps

灯丝电压　filament voltage

灯座　lamp holder

等级分界　material specification break

等径三通　straight tee

等离子焊　plasma welding

等熵指数　isentropic exponent

低点　low point

低点标高　low point elevation

低电压继电器　under-voltage relay

低功率因数瓦特表　low power-factor wattmeter

低温用阀　cryogenic service valve

低压　low-voltage/low-tension/low pressure

低压避雷器　low voltage arrester

低压镀锌焊接钢管　low pressure galvanized pipe(l. p. galv.)

低真空阀　low vacuum valve

底板　base plate

底板底面　bottom of base plate

底阀　foot valve

底面/底下　bottom

底平　flat on bottom

底漆　primary coat/bottom coat

地秤室　weigh-bridge room

地脚螺栓　anchor bolt/foundation bolt

地坑/井　pit

地漏　floor drain

地面　ground

地面标高　ground level

地面坡度　ground grade

地坪　grade level

地上　above ground

地下　under ground

点焊　spot welding

电伴热　electrical tracing

电池室　battery room

电除尘器　electrostatic precipitator

电吹尘器　electric dust cleaner

电磁阀　electromagnetic valve

电磁流量计　electromagnetic flowmeter

电磁式继电器　electromagnetic relay

电笛　siren

电动单元组合仪表　packaged electronic instruments

电动阀　valve with electrically motorized operation

电动葫芦　motor hoist/electric hoist/electric block

电动机　motor

电动机堵转电流　locked-rotor motor current

电动机加热器　space heater (for motor)

电动机起动顺序　sequence of motor starting

电动秒表　electric second-meter

电动式继电器　electrodynamic relay

电感负荷　inductive load

电感线圈　induction coil

电工系统图图形符号　graphic symbols for electric system

电弧焊　arc welding

电极　electrode

电极信号　electrode signal

电胶木　bakelite

电接点压力表　electric contact pressure gauge

电接点液位计　remote level indicator

电控箱　electric control panel

电喇叭　horn

电缆编号　No. of line

电缆槽(架)　cable tray(channel)

电缆穿管与管道交叉　cable protective pipe across a pipeline

电缆吊架　cable hanger

电缆敷设图　cable laying diagram

电缆膏　cable compound

电缆沟　electric cable duct/electrical trench

电缆故障探伤仪　cable fault detector

电缆汇线槽　cable duct

电缆联系图　cable hook-up diagram

电缆与热力管道交叉敷设　cable running across heat pipeline

电缆与水管平行　cable running parallel to a water supply pipe

电缆终端头　pothead of cable or cable end

电力变压器　power transformer

电力电压　power voltage

电力平面布置图　electric power layout plan

电力系统图　electric power system diagram

电流　current

电流表　ammeter

电流测量回路　current-measuring circuit

电流互感器　current transformer

电流继电器　current relay

电流密度　current density

电流线圈　current coil

电炉　electric furnace

电气标准图　electric standard drawing

电气管线表　list of wire/cable and conduits

电气盘　electrical panel

电气施工图　electric working drawing

电气图纸目录　contents of electric drawings

电气线路　electric circuit

电容　capacitor

电容式料位计　capacitance level meter

电容式压力变送器　capacitance pressure transmitter

电熔焊　electric fusion welding

电梯　electrical lift；Elevator

电位器　potentiometer

电压　voltage

电压表　voltmeter

电压表转换开关　voltmeter change-over switch

电压等级　voltage grade

电压干线　voltage main line

电压互感器　potential transformer

电压互感器柜　potential transformer cabinet

电压继电器　voltage relay

电压监视　voltage supervision

电压降　voltage drop

电压损失　voltage loss

电源　power Supply

电源配电箱　source distribution cabinet

电源切除　switch off the power supply

电源小母线　power supply miniature bus

电渣焊　electroslag welding

电子秤　electronic weighter

电子继电器　electronic relay

电阻焊　electric resistance welding，ERW

电阻式温度计　resistance thermometer

垫片　gasket

垫片(垫平用)　shim

吊车梁　crane beam/hoist beam

吊耳　ear

吊勾　hook

吊架　hanger

吊梁　hoisting beam

吊装孔　erection opening

掉闸回路　trip circuit

掉闸回路断线信号　breakage signal of trip circuit

掉闸线圈　tripping coil

掉闸音响信号　tripping audible signal

蝶阀　butterfly valve

蝶形螺母　wing nut

丁苯橡胶　styrene-butadiene rubber

丁基橡胶　butyl rubber

丁腈橡胶　nitride butadiene rubber

顶板　top plate

顶面/面　top

顶平　flat on top

定时限继电器　definite time relay

定位销　dowel pin

定子绕组测温回路　temperature measuring circuit for stator winding

动船法　moving boat method

动力配电箱　power distribution cabinet

动圈式显示仪表　moving coil indicator

动态容积法　dynamic gauging

动态质量法　dynamic weighing

斗式提升机　bucket elevator

渡越时间式超声流量计　transit time meter

镀锌　galvanized (galv)

镀锌钢管　galvanized steel pipe

镀锌角钢　galvanized steel angle

镀锌煤气管　galvanized gas pipe

镀锌铁皮　galvanized (sheet) iron

镀锌铁丝　galvanized wire

镀锌铁丝网　galvanized wire mesh

端子出线　terminal outgoing

端子接线图　diagram of terminal connections

端子排　terminal board

端子箱　terminal box

短半径弯头　short radius elbow, SRE

短节(短管)　nipple, NIP

短路电流　short-circuit current

短路电流最大有效值　maximum effective value of short-circuit current

断接卡　connecting clamp

断路器　circuit breaker

断路器事故掉闸信号　fault trip signal of breaker

断面系数　section modules

断相保护　phase failure protection

锻钢　forged steel

锻件　forging

锻造　forging

对焊　butt welding

对焊的　butt welded

对焊法兰　weld neck flange

对焊连接　butt welded joint

对中心/找正　alignment

多点温度巡测仪　multi-point logger

多回路的　multi-loop

多级泵　stage pump/multistage pump

多级压缩机　multiple stages compressor

多量程仪用电流互感器　multi-range current transformer for measurement

多路阀　multiport valve

多普勒超声波流量计　Doppler meter

多普勒效应　Doppler effect

多切点切换开关　multi-point change-over switch

多声道斜束式超声波流量计　multi-path diagonal-beam flowmeter

多隙避雷器　multigap arrester

多相流　multiphase flow

多种电源插销箱　receptacle box for miscellaneous power supplies

E e

额定电流　rated current

额定电压　rated voltage

额定值　rating

颚式破碎机　jaw breaker/jaw crusher

二次电压　secondary voltage

二次装置　secondary device

二阀组　2-valve manifold

二线制　two-wire system

F f

发电机　generator

发电机与工作母线并车　synchronization of generator to working bus

阀门及配件制造工业标准化协会(美) manufacturers' Standardization Society of Valve

阀形避雷器　auto-valve arrester

法兰的　flanged

法兰盖　blind flange

法兰连接　flanged joint

法兰面　flange facing/facing of flange

法兰取压孔　flange pressure tapping

翻板阀　flap valve

反馈回路　feedback loop

反力　reaction

反射内存　reflective memory

反应器　reactor

反应式电动机　reaction motor

方锉　square file

方垫圈　square washer

方钢　square bar

方螺母　square washer

方头螺栓　square head bolt

方位　orientation

防爆电气设备　explosion-protected electrical apparatus

防爆铠装电缆密封接头　ex(d). armored-cable packing gland

防爆控制按钮　explosion-proof control push-button

防爆密封接头　ex(d) packing gland

防爆密封接头挠性管　flexible conduit with ex(d). packing gland

防爆照明灯开关　explosion-proof lighting switch

防腐油　anti-corrosive oil

防护式灯开关　protective switch

防护式开关　guard type switch

防火间距　fire protection spacing

防雷接地　grounding for lightning

防水防尘灯　water and dust proof lighting fitting

防水铠装电缆密封接头　water-proof armored-cable packing gland

防水密封接头　water-proof packing gland

防水密封接头挠性管　flexible conduit with water-proof packing gland

防水挠性管　water-proof flexible conduit

防锈漆　antirust paint

仿真机　simulator

放大图　enlarged view

放空　vent

放空阀　vent valve

放空丝堵　vent plug

非对称性指数　index of asymmetry

非金属包覆垫片　non-metallic jacked gasket

非铺砌区　unpaving area

非线性　nonlinearity

非淹没流　modular flow

非淹没限　modular limit

废品　discard

分贝　decibel

分辨力　resolution

分段屏　sectionalizing panel

分离器　separator

分离容器　separator or seal chamber

分励　shunt trip

分流器　shunt

分散控制系统　distributed control system, DCS

分析设计阶段　analytical engineering phase

分析室　analyzer

分子泵　molecular pump

酚醛漆　phenolic paint

粉碎机/磨碎机　grinder/pulverizer/disintegrator/atomizer

风机　draught Fan

风冷式　air-cooled

风扇变速开关　fan speed regulator switch

风压　air pressure

封离真空装置　sealed vacuum device

蜂鸣器　buzzer

弗劳德数　Fronde Number

浮充屏　floating panel

浮顶罐　floating roof tank;pontoon roof tank

浮阀塔　valve tower

浮力修正　buoyancy correction

浮球式视镜　floating ball sight flow indicator

浮子　float/sinker

浮子式液位计　float level indicator

符号　symbol

辅助电源　auxiliary source

负荷计算　load calculation

负极　negative pole

负载阻抗　load impedance

附壁效应　coanda effect

附件　accessory

附属设备/附件　attachment equipment

复合管　clad pipe

复合压缩机/串联压缩机　combined compressor/combination compressor

复用图　reproducible drawing

G g

概略版设备布置图　conceptual plot plan

干扰　disturbance

干式气柜　dry gas-holder

干线　main line

干燥器　dryer

感应调压器　induction voltage regulator

感应电动机　induction motor

感应电流　induced current

感应雷击　induction lightning stroke

钢板　plate

钢带　strap steel

钢管　steel pipe

钢号码　steel mark

钢结构　steel structure

钢结构顶面　top of steel

钢锯　hacksaw

钢卷尺　steel band tepe

钢丝钳　cutting pliers

钢丝绳　steel wire rope

钢直尺　steel rule

钢制电缆管　rigid steel conduit

高点　high point

高度　altitude

高度游标卡尺　altitude caliper vernier

高硅铸铁　high silicon cast iron

高空环境模拟舱　high-altitude environment simulation cabin

高频淬火　high-frequency quenching

高斯求积法　Gaussian integration method

高压　high-voltage/high-tension

高压负荷开关　H. V. load break switch

高压隔离开关　H. V. disconnecting switch

高压开关柜　H. V. switchgear

高压配电室　H. V. distribution room

高压试验变压器　H. V. testing transformer

高压水银灯　high pressure mercury vapor lighting fitting

高压水银荧光灯　H. P. mercury fluorescent lighting fitting

高真空阀　high vacuum valve

隔爆灯　explosion-proof lighting fitting

隔断阀　isolating valve

隔离变压器　isolating transformer

隔离容器　seal chamber

隔膜阀　diaphragm valve

隔热/隔离/绝缘　insulation

隔热分界　insulation break

隔音　sound insulation

给定值　set point

给水流量　feedwater flow

根部阀　root valve/primary valve/header valve

更衣室　locker room

工厂(车间)焊接　shop weld

工厂/装置　plant

工具钢　tool steel

工控机　industrial personal computer

工频耐压　high-voltage test with working frequency

工艺故障　fault in process

工艺控制图　process control diagram

工艺流程图　process flow diagram

工艺位号　process item No.

工字钢(工字梁)　I-beam

工作/保安电源切换　transfer of working and emergency power supply

工作程序　working procedure

工作电压　working voltage

工作电源　working power supply (electric source)

工作电源分合闸　on and off of working electric source

工作电源失电报警　no-voltage alarm of working electric source

工作电源投入　throw-in of working electric source

工作接地　working grounding/working earthing

工作量器　calibrated measuring (volumetric) tank

工作条件　working conditions

工作/操作温度　working temperature/operating temperature

工作压力　working pressure

弓形夹　cramp frame

公称流量　nominal flow rate

公称压力　nominal pressure

公称直径　nominal diameter

公用工程流程图　utility flow diagram, UFD

公章　official seal

公制螺纹　metric thread

功耗　electrical power consumption

功率继电器　power relay

功率因数　power factor

功率因数表　power factor meter

供电电压　supply voltage

供电盘　power supply box

共模电压　common mode voltage

共模干扰　common mode interference

共模信号　common mode signal

共模抑制比　common mode rejection ratio

共振　resonance

鼓风机　blower

固定点　fix point/anchor point

固定电阻(器)　fixed resistance

固定顶罐　stationary roof tank

固定架　anchor

固定检速架　stationary array

固态继电器　solid state relay

故障电流　fault current

故障时关闭　valve closes on failure of actuating energy

故障时开启　valve opens on failure of actuating energy

顾客　client/customer

关/闭阀回路　circuit for closing valve

关闭后加铅封/封关　car seal close

关闭后加锁/锁关　locked closed

关联设备　associated apparatus

观察窗　viewing window

管壁取压孔　wall (pressure) tapping

管长　conduit length

管道　piping

管道泵　inline pump

管道布置　piping assembly/piping layout

管道布置平面　piping arrangement plan

管道材料规定　piping material specification

管道等级　piping class

管道附件　piping attachment

管道及仪表流程图　Piping and Instrument diagram

管道跨距　line span

管道研究　piping study

管道元件　piping element

管底　bottom of pipe

管顶　top of pipe

管堵(丝堵)　plug

管段/短管　spool piece/spool

管墩/低管架　sleeper

管沟　pipe trench

管箍　coupling

管架　piping support

管间距　line spacing

管件直连　fitting to fitting

管接头(管箍)　coupling/full coupling

管卡　clamp

管壳　shell

管口　nozzle

管口方位　nozzle orientation

管廊　pipe rack

管流　pipe flow/duct flow

管螺纹　pipe thread

管螺纹连接　pipe threaded joint

管帽　pipe cap
管内底　invert/inside bottom of pipe
管内底标高　invert elevation
管式熔断器　cartridge fuse
管托　shoe
管形避雷器　tubular arrester
管子(标准规格)　pipe
管子(非标规格)　tube
管子内螺纹　internal pipe thread
管子钳　pipe wrench
惯性矩　moment of inertia
盥洗室/厕所　closet/toilet/washroom
罐/鼓/筒/桶　drum
罐底排污阀　flush-bottom tank valve
光电继电器　photoelectric relay
光滑突面　smooth raised face
光面/光滑面　plain face
光纤　fiber
广照型工厂灯　wide lit type industrial fitting
规划布置阶段　planning stage
规划研究版设计布置图　planning plot plan
规则速度分布　regular velocity distribution
硅酸钙　calcium silicate
硅酸铝纤维　aluminasilicate fibre
硅有机树脂　silicone
硅藻土　kieselguhr/diatomite
硅整流器　silicon rectifier
滚动支架　rolling hanger

锅炉 boiler

锅炉跟踪方式 boiler following mode

锅炉压力 boiler pressure

国际标准化组织 International Standardization Organization,ISO

国际单位制 System International

国际电工技术委员会 International Electrotechnical Commission,IEC

过/欠电压 over/under voltage

过程变量 process variable

过电流 over current

过电流保护回路 over-current protective circuit

过电流继电器 over-current relay

过电压继电器 over-voltage relay

过渡接头 adapter

过滤器 filter

过氯乙烯漆 ethylene perchloride paint

过载电流表 overload ammeter

H h

含湿量 humidity ratio

焊接 welding

焊接钢管 welded steel pipe

焊接管帽 welded cap

焊接件 weldment

焊丝 welding wire

焊条 welding electrode(rod)

毫安表 milliammeter

合格品 accepted product/conforming article

合金钢 alloy steel

合金钢管　alloy steel pipe

合金结构钢　structural alloy steel

合同版设备布置图　award plot plan

合同号　contract number/Cont. No.

合闸电源母线　closing power source bus

合闸回路　closing circuit

合闸位置继电器　close position relay

合闸线圈　close coil

河床坡度　bed slope/bottom slope

核辐射式料位计　radiation level meter

荷载　load

黑色金属　ferrous metal

恒定平均流量的脉动流　pulsating flow of mean constant flow rate

恒力吊架　constant hanger

恒温箱　constant temperature box

桁架　girder

横截面　intersecting surface

横截面内的平均动压　mean dynamic pressure in a cross-section

红色障碍灯　red obstruction lamp for aviation

红外线　infrared

虹吸管　syphon/siphon

喉部　throat

厚度　thickness

呼吸阀　breather valve

互相连接的导线　cross connection of wires

护线帽　bushing

华氏　Fahrenheit

滑触线　trolley conductor

滑动架 sliding support

滑石粉 talc powder

滑套法兰/平焊法兰 slip-on-welding flange

滑线变阻器 sliding rheostat

划针 scriber

环境温度 ambient temperature

环境压力 ambient pressure

环连接面/梯型槽密封面 ring joint face

环室 annular chamber

环头螺栓 Eye bolt

环形喉部临界流文丘里喷嘴 toroidal throat Venturi nozzle

环形间隙 annular space

环氧树脂 Epoxy resin

环氧树脂漆 Epoxy resin paint

缓冲器 knock-out drum

换向器 diverter

黄铜 brass

灰铸铁 grey cast iron

回火 tempering

回路 loop

回路编号 circuit No.

回路偏差 loop error

混合罐 mixing tank(drum)

混合孔板 mixing orifice

混合器 mixer

混流泵 mixed flow pump

混凝土顶面 top of concrete

活扳手 adjustable spanner

活接头　union

活塞式压力表　piston type pressure gauge

活塞压缩机　crankshaft piston compressor

火炬　flare

霍尔压力变送器　Hall pressure transmitter

J j

击穿保护器　puncturing safety device

击穿电压　puncture voltage

击穿强度　puncture intensity

击穿熔断器　puncture lightning arrester

机械加工　machining?

机用铰刀　machine reamer

机组负荷指令　unit load command

积点法　point method

积分法　integration method

基础　foundation

基础设计　basic design

基础压力　base pressure

激磁电压　exciting voltage

激磁回路　exciting circuit

激光焊　laser beam welding

极限压力　ultimate pressure

极限有效电流　limit effective current

棘轮扳手　ratchet wrench

计量　metrology

计量泵/定量泵　metering pump

计量检定　metrological verification

计量鉴定证书　metrological expertise certificate

计量确认　metrological confirmation

计量装置　calculating device

计算电流　calculating current

计算电压　calculating voltage

计算器　calculator

继电器　relay

继电器常闭触点　Relay N. C. contact

继电器常开触点　Relay N. O. contact

加厚　thicken

加拿大标准协会　Canadian Standard Association，CAS

加强板　stiffener

加权算术平均值　weighted arithmetic mean

加热器　heater

夹持环　clamp ring

夹紧式胶管阀　pinch valve

夹套阀　jacketed valve

夹套管　jacketed tracing

夹装式超声波流量计　clamp-on flowmeter

架顶面　top of support

尖嘴钳　pointed tongs

间断焊　intermittent welding

监控　supervision

监视器　monitor

减压阀　reducing valve，RV

减压器　reducer

减振器　snubber

检测点　measuring point

检测仪表　detecting and measuring instrument

检测元件/传感元件　sensor

检定　verification

检定规程　regulation for verification

检定证书　verification certificate

检验　inspection

建立时间　settling time

建设　construction

建筑北　construction north

建筑物　building

渐近速度系数　velocity of approach factor

鉴别阈　discrimination threshold

降压变压器　step-down transformer

交/直流　alternating/direct current

交换机　switch

交货付款　cash on delivery

交流低压配电屏　ACLV switchboard (distribution panel)

交流电的相别　AC phase sequence

交流电焊机　AC welding machine

交流电源　AC power supply

交流电子稳压器　AC electronic voltage stabilizer

交流配电线路　AC distribution circuit

交流异步电动机　AC asynchronous motor

交直流两用钳型电流表　AC/DC multi-purpose tongtester

角阀/角式截止阀　angle valve

角钢　angle steel

角焊　fillet welding

角接取压孔　corner pressure tapping

角位移　angular rotation

角形避雷器　horn arrester

搅拌器　agitator

校核　checked

校正系数　correction factor

校准　calibration

阶跃响应时间　step response time

接触器　contactor

接待室/会客室　reception room

接地保护　ground protection/earth protection

接地电路　earth（ground）circuit

接地电阻　earth resistance

接地电阻测量仪　earthing resistance tester

接地干线　grounding main line

接地故障　ground fault/earth fault

接地回路　earthed circuit

接地或接零线路　grounding or neutralizing circuit

接地极　earth electrode/pole

接地继电器　earthing relay

接地铜线　grounding copper wire

接地网　grounding network/earthing network

接地系统　grounding system/earthing system

接地线　ground connector

接地信号回路　grounding signal circuit

接地装置　grounding device/earthing device

接管式防爆密封接头　ex(d). packing gland for connecting pipe

接管式防水密封接头　water-proof packing gland for connecting pipe

接口　interface

接闪装置/避雷器　lightning arrester

接线端子　terminal

接线盒/箱　junction box

接续图　continue on drawing

接续线　match line

节流阀　throttle valve

节流孔　orifice

节流装置　throttle devices

结晶器　crystallizer

截面(积)　cross section area

截尾文丘里管　truncated Venturi tube

截止阀(球心阀)　globe valve

介质损失角　dielectric loss angle

界流　critical flow

界区条件　battery limit condition

金属包覆垫片　metallic jacket gasket

金属垫片　metallic gasket

金属软管　metal hose

进气阀　gas admittance valve

进气系统　gas admittance system

进线屏　incoming line panel

经典文丘里管　classical Venturi tube

晶体管　transistorized diode

精加工　finishing

精馏塔　fractionating tower

精密度　precision

精制的　finished/fine

净正吸入压头　net positive suction head

径向跳动　radial run-out

静电电容器柜　static capacitor cabinet

静电感应　electrostatic induction

静电接地　static grounding

静态容积法　static gauging

静态质量法　static weighing

静压　static pressure

静压皮托管　static pressure Pitot tube

静压取压孔　static pressure tapping

酒精温度计　alcohol thermometer

就地盘　local panel

就地仪表　local instrument

局部照明变压器　local lighting transformer

局部照明灯　local lighting fitting

局域网　local area network，LAN

距离　distance

锯片铣刀　saw web cutter

聚氨酯　Polyurethane

聚氨酯漆　Polyurethane paint

聚苯乙烯　Polystyrene

聚丙烯　Polypropylene

聚丙烯外壳　Polyacrylic cover

聚氯乙烯　Polyvinyl chloride，PVC

聚三氟氯乙烯　Poly chlorotrifluoroethylene，PCTFE

聚四氟乙烯　Polytetrafluoroethlene

聚四氟乙烯滑动板　PTFE sliding plate

聚碳酸酯　Polycarbonate

聚乙烯　Polyethylene

绝对标高　absolute elevation

绝对压力　absolute pressure

绝缘包布　insulating tape

绝缘电线　insulated wire

绝缘膏　insulating compound

绝缘监视　insulation supervision

绝缘漆　insulating varnish

均压环　piezometer ring

K　k

卡套式　ferrule-type

开度计　opening meter

开度计回路　circuit for opening meter

开阀回路　circuit for opening valve

开工会议　kick-off meeting

开关放大器　switch amplifier

开关箱　switch box

开关型号　switch type

开环控制　open loop control

开口三角形连接的三相绕组　3-phase winding with open delta connection

开口销　cotter pin

开启后加铅封/封开　car seal open

开启后加锁/锁开　locked closed

科氏力质量流量计　coriolis mass flowmeter

可编程逻辑控制器　programmable logic controller, PLC

可拆短管　removable joint spool

可拆卸的电气连接　removable connection

可拆卸的真空封接　demountable joint

可锻铸铁　malleable iron

可控硅　silicon controlled rectifier

可控硅励磁装置　silicon controlled rectifier excitation device

可控硅整流箱屏　silicon controlled rectifier box panel

可膨胀性/膨胀系数　expansibility/expansion factor

空冷器　air cooler

空气电加热器/电炉 electric heating furnace

空气断路器　air circuit-breaker

空气断路器电机操作机构　motor operating mechanism for air circuit-breaker

空气净化系统　air purification system

空气压缩机　air compressor

孔/洞　hole

孔板　orifice plate

孔板法兰　orifice flange

控制　control

控制按钮　control push-button

控制电源　control supply

控制电源开关　switch for control supply

控制电源中间继电器　auxiliary relay for control supply

控制掉闸　control trip

控制阀/调节阀 control valve

控制回路　control circuit

控制器　controller

控制室　control room

控制台　console

控制线路　control circuit

控制箱　control cabinet

控制箱面部布置图　surface arrangement of control box

控制箱台面展开图　surface developed diagram of console

控制仪表　control instrument

控制原理图　principle control diagram

控制装置　control device

控置室　control room

跨度　span

跨接线　jumper

快闭阀　quick closing valve

快速接头　quick coupling

宽度　width

矿渣棉　mineral wool

馈路　feeder circuit

扩孔　hole expanding

扩散硅压力变送器　diffused silicon pressure transmitter

L l

蜡　wax

雷电概率　lightning and thunder probability

雷电日　thunderbolt days

雷击　lightning stroke

冷端补偿器　cold junction compensator

冷端温度　cold end temperature

冷紧/冷拉　cold spring

冷凝容器　condensate pot

冷凝弯/圈　pipe syphon

冷却器　condenser

冷却塔　cooling tower
离心泵　centrifugal pump
离心分离机　centrifugal separator
离心过滤机　centrifugal filter
离心机　centrifuger
离心压缩机　centrifugal compressor
力矩　moment of force
历史趋势画面　historical trend display
立杆弯灯　goose-neck post lamp/pole lamp
立式泵　vertical pump
立式的/垂直的　vertical
励磁发电机　excitation generator
励磁故障　fault in excitation
励磁机　exciter
沥青　asphalt
沥青漆　bituminous varnish
连接/接头　connection
连接板　tie plate
连接杆　tie rod
连接片　connecting link
连接器　connector
连接条　connecting strip
连接箱　junction box
连续处理真空设备　continuous treatment vacuum plant
连续负荷　continuous load
联锁　interlocking
联锁继电器　interlock relay
联锁解除　release of interlock

联锁开关　interlock switch

梁　beam

梁顶面　top of beam

量程　span

量热式质量流量计　thermal mass flow meter

临界电流　critical current

临界电压　critical voltage

临界截面　throat

临界流函数　critical flow function

临界流流量计　critical flowmeter

临界流喷嘴　critical nozzle

临界流文丘里喷嘴　critical Venturi nozzle

临界压力　critical pressure

临界压力比　critical pressure ratio

临时粗滤器　temporary strainer

灵敏度　sensitivity

零点压缩　zero suppression

零母线　neutral bus

流出系数　discharge coefficient

流动剖面　flow profile

流量变送器　flow transmitter

流量范围　flow rate range

流量管/测量管　meter tube

流量计　flowmeter

流量计误差特性曲线　error performance curve of flowmeter

流量开关　flow switch

流量信号　flow signal

流速计　current meter

流体单元动压　dynamic pressure of fluid element
流体的绝对静压　absolute static pressure of the fluid
六角螺母　hexagonal nut
六角头螺栓　hexagonal head bolt
楼面　floor
漏斗　funnel
露点温度　dew point temperature
露点仪　dew-point meter
炉膛安全监控系统　furnace safeguard supervisory system，FSSS
炉子　furnace
路灯　street lamp
路由器　router
罗茨泵　the roots vacuum pump
逻辑元件　logical element
螺杆泵　screw pump
螺杆压缩机　helical screw compressor
螺栓冷紧　bolt cold tightening
螺栓热紧　bolt hot tightening
螺栓圆　bolt circle
螺栓圆直径　bolt circle diameter
螺纹的　threaded
螺纹法兰　threaded flange
螺纹管帽　threaded cap
螺纹焊管座　threaded let
螺纹连接　threaded joint
螺旋板换热器　spiral plate exchanger
螺旋式气柜　helical gas-holder
螺旋输送机　screw conveyer/spiral conveyer

落差　fall

铝　aluminum

铝板　aluminum sheet

铝避雷器　aluminum cell arrester

铝镁合金　aluminum magnesium

氯丁橡胶　neoprene rubber

氯化聚氯乙烯　chlorinated ploy vinyl chloride

氯化聚醚　chlorinated polyether

氯化聚乙烯　chlorinated poly ethylene, CPE

氯磺化聚乙烯　chlorosulfonated polyethylene

M　m

马氏体不锈钢　Martensitic stainless steel

买方(供货)　by buyer

卖方(供货)　by seller

脉动电压　pulsating voltage

满刻度流量　full scale flow rate

慢扫描示波器　slow scanning oscillograph

盲板/管道盲板　blank/blind/line blank

盲孔　blind hole

锚固件　clip

铆钉　rivet

铆接　riveting

美国标准直管螺纹　American standard taper pipe thread

美国材料试验协会　American Society for Testing and Materials, ASTM

美国国家标准协会　American National Standard Institute, ANSI

美国机械工程师学会　American Society of Mechanical Engineers, ASME

美国石油学会　American Petroleum Institute, API

美国仪表学会　Instrument Society of American, ISA

门　gate

蒙乃尔合金(镍铜合金)　Monel

迷宫压缩机　labyrinth compressor

密闭照明灯开关　hermetic lighting switch

密封接头　sealing fitting

面积　area

面漆　finishing coat

面至面　face to face

临界流　critical flow

明沟　open trench

明火地点　open fire place

明渠流　open channel flow

明装单极板钮开关　surface-mounted single-pole toggle switch

明装双极插座　surface-mounted 2-pole receptacle

铭牌　name plate

铭牌框注字　name plate denotation (inscription)

命名表,管道表　nomenclature

模糊控制　fuzzy control

模拟报警信号　semigraph and alarm signal

模拟信号　analogue signal

模拟信号触点　contact for semigraph signal

模数转换　analog-to-digital conversion

模压件　molded part

膜盒　bellows

膜盒式压力表　bellows-gauge

膜片　diaphragm

膜片压缩机　diaphragm compressor

磨削　grinding

莫氏锥柄铰刀　morse taper reamer

母排　busbar

母线　Bus-bar

母线槽　busway

母线接地　bus ground

母线截面　section of bus

母线联络柜　bus tie cabinet

木锤　wooden mallet

目标负荷　target load

N　n

耐电压　breakdown voltage

耐火砖　fire brick

耐油橡胶管　oil-proof rubber tubes

挠性管(柔性管)　flexible hose

内部接线图　internal (inside) wiring diagram

内电阻　internal resistance

内浮顶罐　floating internal roof tank

内环　inner ring

内径百分表　internal centigrade gauge

内六角扳手　inner hexagon wrench

内螺纹　female thread

内啮合齿轮泵　internal gear rotary pump/Crescent gear pump

能量　energy

能自动返回的常闭按钮触点　self-return button with N. C. contact

泥浆泵　slurry pump/sludge pump/Mud pump

逆电流　inverse current

逆流继电器　reverse-current relay

扭矩　torque

O o

欧姆表　Ohm meter

P p

排出阀　discharge valve

排出口　exhaust

排放丝堵　drain plug

排气孔　vent holes

排气量　displacement capacity

排污阀　blowdown valve

排泄孔　drain holes

排液　drain

排液阀　drain valve

排液漏斗　drain funnel

盘式过滤机　tray filter

旁路阀　by-pass valve

旁通管路　by-pass line

抛光　polishing/buffing

刨削　planning shaping

泡沫玻璃　foam glass/cellular glass

泡沫混凝土　foamed concrete/cellular concrete

泡沫聚苯乙烯　expanded poly styrene，EPS

泡沫站　foam station

配电柜　power distribution cabinet

配电室　substation

配电箱　distribution cabinet

配对温度传感器　temperature sensor pair

配件　fitting part

配照型灯　standard dome lighting fitting

喷射器　ejector

喷头　sprayer

膨胀节(伸缩接头)　expansion joint

膨胀螺栓　expansion bolt

膨胀系数　expansion factor

皮托管　pitot tube

匹配式玻璃金属封接　matched glass-to-metal seal

偏差　deviation

偏流测向探头　yaw probe

偏心度　eccentricity

偏心孔板　eccentric orifice plate

偏心异径管　eccentric reducer

偏置调节　bias adjustment

漂移　drift

频率表　frequency meter

频敏电阻器　frequency sensitive rheostat

频谱　spectrum

平焊　flat welding

平焊法兰　slip on flange/slip on weld flange

平滑性　flatness

平均粗糙度　mean roughness

平均温度　average temperature

平均轴向流体速度　mean axial fluid velocity

平均轴向流体速度点　points of mean axial fluid velocity

平面度　flatness

平台　platform/flat/floor/surface plate

平行　parallel

平行度　parallelism

平行夹　parallel frame

屏蔽泵　shielding can-type pump/canned pump

屏蔽接地　screen earthing/shielding ground

屏编号　panel No.

瓶车箱　synchronizing cabinet

坡度　slope/Grade

坡口　groove

破碎机　breaker/cracker mill/crusher/crushing mill

剖面图　section

铺砌区　paving area

Q q

起动设备　starting device

起点　starting

起动按钮　starting push-button

起动器　starter

起始次暂态短路电流　initial subtransient short-circuit current

起重机/吊车　crane

气动泵　pneumatic pump

气动单元组合仪表　packaged pneumatic instruments

气动阀　pneumatic operated valve

气动管　pneumatic tube

气动基地式调节仪表　pneumatic local control device

气柜　gas bolder

气焊　gas welding

气密式组合开关　hermetic packet type switch

气熔焊　fusion gas welding

气体过滤器　filter

气体流量标准装置　gas flow standard facilities

气压继电器　gas-pressure relay

气压试验　pneumatic testing

气液分离器　gas-liquid separation

气源管道　air piping

汽机跟踪方式　turbine following mode

汽机监视仪表　turbine supervisory instrument

汽轮机　steam turbine

汽压　steam pressure

千分表　clock gauge

千分表架　clock gauge frame

千分尺　micrometer

铅　lead

前级真空阀　backing valve

前级真空管路　backing line

前级真空容器/贮气罐　backing reservoir

前缘　leading edge

钳加工　bench work

潜水泵　submerged pump;immersible pump

嵌入式荧光灯　flush type fluorescent lighting fitting

强励磁接点　shock excitation contact

强制周期检定　mandatory periodic verification

墙　wall

墙架　support on wall

切断电流　rupturing current

切断阀　block valve/shut-off valve/stop valve

切换片　transfering link

切线　tangent line

倾角　angle of inclination

倾斜度　angularity

清除口/清洗口　clean out

球阀　ball valve

球罐　spheroid;Spherical tank

球墨铸铁　nodular cast iron/nodular cast graphite iron

球型避雷器　spherical arrester

球型补偿器　ball type expansion joint

取压点　pressure measuring/tapping point

取压管接头　pressure tap

取压孔　pressure tapping/pressure taps

取样阀　sampling valve

取样接口　sampling connection

取样冷却器　sample cooler

取源部件　tap

去毛刺　deburring

全国电气制造商协会(美)　National Electrical Manufactures Association，NEMA

全国消防协会(美)　National Fire Protection Association，NFPA

全平面/满平面　flat face/full face

全视镜　full view sight flow indicator

全跳动　total run-out

裙座　skirt

R r

燃料安全系统　fuel safety system

燃烧器管理系统　burner management system

燃烧器控制系统　burner control system

扰动　upset/disturbance

绕线式电动机　wound-rotor induction motor

热处理　heat treatment

热电偶　thermocouple

热电阻　thermal resistance

热继电器　thermal relay

热能表　heat meters

热膨胀　thermal expansion

热水伴管　hot-water tracing

热元件　thermal element

人工干预　manual intervention

人工接地　artificial grounding

人机接口装置　man-machine interface device

人孔　manhole

人身保护　personal protection

日本工业标准　Japanese Industrial Standard，JIS

容积式流量计　positive displacement flowmeter

熔断电流　blow-out current

熔断器　fuse

熔融金属真空封接　molten metal seal

熔丝　fuse-link

肉眼检验/外观检验　visual testing

入口　inlet/suction

入口温度　inlet temperature

入口压力　inlet pressure

软管接头　hose connection

软管站　hose station

软手操　soft manual operation

S s

三阀组　3-valve manifold

三极高压断路器　3-pole HV circuit-breaker

三角锉　triangular file

三角刮刀　triangular scraper

三脚架　triangular support

三路开关　3-way switch

三通　tee

三通阀　3-way valve

三通球阀 3-way ball valve

三通旋塞阀　3-way cock valve

三相滑环感应电动机　3-phase slip-ring induction motor

三相三绕组变压器　3-phase tertiary winding transformer

三相三线有功电能表　3-phase three-wire kilowatt-hour meter

三相笼型感应电动机　3-phase squirrel-cage induction motor

三相四线制标准电能表　3-phase 4-wire standard watthour-meter

三相电能表　3-phase watt meter

三相无功电能表　3-phase kilovar-hour meter

三相自耦变压器　3-phase auto-transformer

散发火花地点　send-out spark place

扫描速率　scan rate

砂轮机　emery wheel grinder

闪光母线　flashing-bus

闪光小母线　flashing miniature bus

闪光信号回路　flashing-signal circuit

闪光装置　flashing device

上升时间　rise time

上位计算机　upper computer

上下限值　high/low limit

设备布置　equipment arrangement/equipment layout

设备材料表　list of equipment and materials

设备容量　consumer capacity

设计北　design north

设计规定　design specification

设计规定汇总表　design specification summary sheet，DSSS

设计温度　design temperature

设计文件　design document

设计压力　design pressure

设计注释　design note

摄氏　centigrade

深度　depth

深度游标卡尺　depth caliper vernier

深井泵　deep well pump

深照型灯具　high bay lighting fitting

审定　approved

审核提纲　check list

渗氮　nitrogen case-hardening

渗碳　carbonization

生产设备　production equipment

声环法　sing around method

声匹配层 acoustic matching layer

声束偏转式超声波流量计 beam deflection flowmeter

失真 distortion

施工版设备布置图 construction plot plan

湿度计 hygrometer

湿式气柜 wet gas holder

十字形螺钉旋具 cruciform screwdriver

石棉 asbest

石棉板 asbestos board

石棉布 asbestos cloth

石墨换热器 graphite heat exchanger

石油管线协会(美) Association of Oil Pipe Lines

时间常数 time constant

时间继电器 time relay

实测垂线平均流速 measured mean velocity on a vertical

实际压力 actual pressure

实时趋势画面 real-time trend display

示波器 oscilloscope

示值 indication

示踪法 tracer methods

事故紧急按钮 emergency stopping push-button

事故信号小母线 miniature bus for fault signal

事故照明灯 emergency lighting fitting

事故照明线路 emergency lighting circuit

视镜 sight glass/sight flow glass

视孔灯 inspection hole lamp

试车 commissioning

试件 testing part

试验　testing

室内温度　indoor temperature

室外地坪　outdoor grade

室外温度　outdoor temperature

释放线圈　releasing coil

手电钻　electric hand drill

手动阀　manually operated valve

手动控制　manual control

手动控制阀　hand control valve

手动跳闸　manual trip

手工铰刀　hand reamer

手虎钳　hang vice

手孔　handhole

手摇泵　hand pump/wobble pump

手摇钻　brace bit

首次检定　initial verification

疏水器　steam trap

输出信号　output signal

输出阻抗　output impedance

输入回路　input loop

输入阻抗　input impedance

输送泵　transfer pump/delivery pump

数据　database

数据表　data sheet

数据采集　data acquisition

数据采集系统　data acquisition system

数据处理　data processing

数据记录　data record/data logging

数据通信系统　data communication system

数控加工　numerical control machining

数字控制　digital control

数字式电液控制　digital electro-hydraulic control system，DEH

数字信号　digital signal

衰减　attenuation

双/单侧支架电缆沟　trench with rack on both sides/one side

双电压电动机　dual-voltage motor

双极带接地插座　2-pole receptacle with grounding contact

双极铁壳开关　2-pole iron-clad switch

双金属温度计　bimetallic thermometer

双色液位计　double coloured indicator

双头扳手　double-end spanner

双头螺栓　stud bolt

双作用往复泵　double action reciprocating pump

水环泵　water ring vacuum pump

水环式真空泵/压缩机　water ring vacuum pump/compressor

水冷却器　water cooler

水冷式　water cooled

水煤气钢管　water-gas steel pipe

水平安装　horizontal installation

水平的/卧式的　horizontal

水压试验　hydraulic testing

水银温度计　mercury thermometer

顺丁橡胶　cis-polybutadiene rubber

顺序控制系统　sequence control system

瞬时　instantaneous

瞬时过冲　transient overshoot

瞬态　transient

说明　description

丝堵　plug

丝锥　screw tap

丝锥扳手　tap wrench

斯特罗哈尔数　Strouhal number

死区　dead band

四通/十字通　cross

伺服电动机　servo-motor

松紧节/花兰螺栓　turnbuckle

松套板式法兰　loose plate flange

松套法兰　lap joint flange

送审版设备布置图　approval plot plan

速动继电器　quick acting relay

速度分布　velocity distribution

速度面积法　velocity area methods

速断及过流断通　instantaneous trip and over-current off/on

塑料　plastic

塑料管　plastic pipe

塑料绝缘线　plastics insulated wire

算术平均值　arithmetic mean

随机误差　random error

榫槽面　tongue groove face

榫面　tongue face

缩流取压孔　vena contract pressure tapping

索赔　claim indemnity

索引图　key plan

锁紧螺母　lock nut

T t

塔 tower

台虎钳 anvil vice

台钻 bench drilling machine

钛 titanium

弹簧垫圈 spring washer

弹簧吊架 spring hanger

弹簧管压力表 bourdon tube pressure gauge

弹簧架 spring support

弹簧托 resting type spring support

碳钢 carbon steel

碳钢管 carbon pipe

碳纤维 carbon fiber

搪瓷 porcelain enamel

搪瓷深照型灯 enameled high bay lighting fitting

陶瓷 ceramic

陶瓷泵 ceramic pump

陶瓷金属封接 ceramic-to-metal seal

陶制泵 stoneware pump

套管 bushing

套管插入深度 jacket insertion depth

套管换热器 double pipe heat exchanger

套间 compartment

套筒扳手 box spanner

特殊管架 special support

特殊失压脱扣器 special no-voltage release

梯形螺纹 trapezoid thread

提升机/卷扬机　hoister/lifter/gig/elevator

体/容积　volume

体电阻率　volume resistivity

体积管　pipe prover

体积流量　volume flowrate

天棚灯　ceiling-mounted lighting fitting

天然橡胶　natural rubber

填充剂　filling agent

填函式补偿器　slip type/packed type expansion joint

填料　filler/packing

填料函　packing box

填料塔　floating distillation tower

调节阀　regulating valve

调节器　regulator

调压变压器　voltage regulating transformer

调压器　voltage regulator

调整　adjustment

调制解调器　modem

调质　quenching and tempering

铁壳开关　metal-clad switch/iron-clad switch

铁素体合金钢管　ferric alloy steel pipe

停车延时回路　delayed shutdown circuit

停止按钮　stopping push-button

通长螺纹/全螺纹　full thread

通风室　ventilation room

通孔　through hole

通信网络　communication network

通用穿线盒　universal fitting

同步电动机　synchronous motor
同步继电器　synchronous relay
同步指示　indicating of synchronism
同相电压　in-phase voltage
同心度　concentricity
同心孔板　concentric orifice plate
同心异径管　concentric reducer
铜　copper
铜锤　copper hammer
投标版设备布置图　proposal plot plan
投光灯　flood-light/projection light
投入指示　indicating of throw-in
凸极同步电动机　salient pole synchronous motor
凸面　male face
突面　raised face
图号　drawing number
图例　legend
涂漆　painting
退火　annealing
托座　stool
脱扣　release/trip
脱扣线圈　trip coil
脱脂　degreasing

W w

瓦斯继电器　gas relay
外部负载　external load
外部接线图　external/outside wiring diagram

外环　outer ring

外汇　foreign exchange

外径　outside Diameter

外壳　(outside)shell/case(shell)/casing/cover

外螺纹　male thread

外协件　teamwork part

弯灯　goose-neck light

弯管　bend

弯矩　bending moment

弯通/弯头　elbow

万能角尺　universal protractor

万用表　avometer

网纹钢板　checkered plate

网状粗滤器　gauge strainer

往复式压缩机　reciprocating compressor

危险区　hazardous area

危险区域　hazardous area

微安表　microammeter

微处理器　microprocessor

微调阀　micro-adjustable valve

维修间　repair room

维修室　maintenance room

位号　No. of location

位移　displacement

温度变送器　temperature transmitter

温度调节仪表　temperature controller

温度开关　temperature switch

温度飘移　temperature drift

温度预告信号　temperature prewarning signal

文丘里管　Venturi tube

文丘里喷嘴　Venturi nozzle

紊流　turbulent flow

稳定流　steady flow

稳定性　stability

稳态　steady-state

稳态短路电流　steady state short-circuit current

涡街流量计　vortex shedding flowmeter

涡轮泵　turbine pump

涡轮流量计　turbine flowmeter

卧式贮罐　horizontal tank

污水坑/污水池　sump pit

无缝　seamless

无缝钢管　seamless steel pipe

无功功率表　reactive power meter/kilovar meter

无机富锌漆　inorganic zinc-rich paint

无损检验　non-destruction testing

无油真空机组　oil free pump system

五阀组　5-valve manifold

误差　error

X　x

吸入阀　suction valve

吸声　sound-absorbing

吸收塔　absorber

洗涤塔　scrubber

洗眼器　eye washer

铣削 milling

系统误差 systematic error

系统增益 system gain

下水管 sewer

先期确认图纸资料 advanced certified final, ACF

显示器 display

显示仪表 display instrument

现场 site

现场按钮 push-button in field

现场焊 field weld

现有回路 existing circuit

线长 wire length

线路/母线/回路 lines/bus/circuits

线路电压 line voltage

线圈 coil

线性 linearity

限定负荷 limited load

限流孔板 restriction orifice

限制杆 limit rod

相别 phase

相对湿度 relative humidity

相位表 phase meter

相序 phase sequence

相移式超声波流量计 phase shift flowmeter

厢式压滤机 box type filer press

箱侧视图 side view of box

箱内框架布置图 arrangement of frames inside the cabinet

详图 detail drawing

详细设计　detail design

详细设计版　detail design issue

项目　project

项目号　job No.

项目审核会　project review meeting

项目情况报告　project status report, PSR

橡胶管　rubber hose

橡套电缆　rubber sheathed cable

消防泵　fire pump

消防栓　fire plug

消音器/消声器　silencer

小母线　miniature bus

协调方式　coordination mode

协调控制系统　coordinated control system

斜垫圈　slant washer

斜盘压缩机　shaft piston compressor

斜坡响应时间　ramp response time

斜束　diagonal beam

携带式交流电桥　portable AC electric bridge

泄漏试验变压器　leakage testing set

卸载阀　unloading valve

信号灯　signal lamp

信号调理　signal conditioning

信号管道　signal piping

信号继电器　signal relay

信号屏　signal panel

信号线路　signal circuit

行程开关　limit switch

型钢　shaped steel

型号　type

型号规格　type specification

性能　feature

休息室　rest room

修改/修订　revision/modify

修正　correction

需要容量　capacity demand

蓄电池屏　battery panel

蓄电池室　battery room

悬臂架　cantilever support

悬索　cable suspension

旋风分离器/旋风除尘器　cyclone/cyclone separator device

旋桨式流速计　propeller type current meter

旋进旋涡流量计　vortex precession flowmeter

旋钮　knob

旋塞　cock

旋塞阀　plug valve

旋涡泵　vortex pump

旋涡流　swirling flow

旋转接头　swivel joint

旋转试验　spin test

询价　inquire

Y y

压差真空系统　differentially pumped vacuum system

压环式　compression-type

压力比　pressure ratio

压力表接头　gauge connector

压力开关　pressure switch

压力损失　pressure loss

压滤机　pressure filter

压缩玻璃金属封接　compression glass-to-metal seal

压缩机室　compressor house/room

压缩因子　compressibility factor

亚临界压力　subcritical pressure

氩弧焊　argon-arc welding

烟囱　stack

延时继电器　time delay relay

延时停车　delayed shutdown

岩棉　rock wool

研磨面搭接封接　ground and lapped seal

验电流器　galvanoscope

氧化锆氧量计　zirconia oxygen indicator

氧化膜避雷器　oxide film arrester

样冲　drift pin

样品　specimen/sample

叶片式真空泵　vane vacuum pump

液环泵/压缩机　liquid ring pump/compressor

液体真空封接　liquid seal

液位变送器　level transmitter

液位计　liquid level meter

液位开关　level switch

液下泵　submerged pump/submersible pump

一次电压　primary voltage

一次装置　primary device

一次装置的校准系数　calibration factor of the primary device

一体式热能表　complete heat meter

一字形螺钉旋具　dash line screwdriver

仪表槽板　instrument trunking

仪表管道　instrumentation piping

仪表盘　instrument panel

仪表试验台　testing stand for instrument

仪表线路　instrumentation line

仪用电感互感器　instrumental voltage transformer

移动软电缆　movable flexible cable

移动式空气压缩机　portable air compressor

移动式压缩机　portable compressor

以太网　ethernet

异径法兰　reducing flange

异径管(大小头)　reducer

异径三通　reducing tee

异径弯头　reducing elbow

阴极射线管　cathode-ray tube

音速喷嘴/临界流喷嘴　sonic nozzle/critical nozzle

音响解除按钮　push-button for sound release

引入线　lead in

引下线　down-lead/down conductor

引下线固定　clamping plate of support for fixing

引下线固定支脚　support for fixing down lead

英国标准　British Standard，BS

荧光灯列　series fluorescent lighting fitting

荧光渗透试验　fluorescent penetrate inspection

硬手操　hard manual operation

永久粗滤器/固定粗滤器　permanent strainer

永久性真空封接　permanent seal

用电设备　electric equipment

用电设备主回路　main circuit of electric equipment

用户变更通知　client change notice

油变阻器　oil immersed rheostat

油浸自冷感应调压器　oil-immersed self-cooled induction voltage regulator

油压千斤顶　hydraulic jack

油毡　asphalt felt

游标卡尺　caliper vernier

有电压的电路　live circuit

有功功率表　active power meter/kilowatt meter

有机玻璃　poly methyl methacrylate，PMMA

有机硅漆　organic silicon paint

有接地极的接地线路网　grounding with grounding electrodes

有色金属　non-ferrous metal

有油真空机组　pump system used oil

有中性点引出线的星形连接的三相绕组　star-connected three phase windings with neutral outlet

右螺纹　right hand thread

预告信号　prewarning signal

预埋件　embedded part/inserted plate

预告母线　prewarning bus

预焊件(设备上)　clip(on equipment)

圆钢　round steel

原材料　raw material

原理图　schematic diagram

圆锉　round file

圆度　roundness

圆球形灯　globe lamp

圆缺孔板　segmental orifice plate

圆筒形喉部文丘里喷嘴　cylindrical throat Venturi nozzle

圆柱形铣刀　cylindrical hobbing

圆嘴钳　round bit tongs

约束　restraint

云母　Mica

允许限值　allowable limit

运转指示　indicating of operation

Z z

载流量　current-carrying capacity

再现性　repeatability

錾子　top chisel

皂膜式气体流量标准装置　standard soap film burette

噪声计　noise meter

增安型　increased-safety

增益　gain

闸阀　gate valve

闸门式真空系统　vacuum system with an air-lock

毡　blanket

展开图　developed diagram

占空比　duty ratio

胀差　differential expansion

兆欧表　Megger/Megohmmeter

照明变压器　lighting transformer

照明电压　lighting voltage

照明干线　lighting main line

照明配电箱　lighting/distribution panel

照明平面布置图　lighting layout plan

照明系统图　lighting system diagram

遮断容量　interrupting capacity

针型阀　needle valve

珍珠岩　perlite

真空保护层　outer chamber

真空窗　vacuum window

真空调节阀　regulating valve

真空断路器　vacuum circuit breaker, VCB

真空阀门的阀座漏气率　leak rate of the vacuum seat

真空阀门的流导　conductance of vacuum valves

真空阀门的特性　characteristic of vacuum valves

真空法兰连接　vacuum flange connection

真空管电压表　vacuum tube voltmeter

真空机组　pump system

真空截止阀　break valve

真空冷凝器/蒸汽冷凝器　device for condensing vapor

真空密封垫　vacuum-tight gasket

真空密封圈　ring gasket

真空平密封垫　flat gasket

真空歧管　vacuum manifold

真空容器的升压速率　rate of pressure rise of a vacuum chamber

真空容器底板　vacuum base plate

真空室　vacuum chamber

真空系统　vacuum system

真空系统的放气速率　degassing/outgassing throughput of a vacuum system
真空系统的漏气速率　leak throughput of a vacuum system
真空系统进气时间　intake time of vacuum system
真空系统时间常数　time constant of a vacuum system
真空限流件　limiting conductance
真空引入线　feedthrough lead through
真空闸室　vacuum air lock
真空钟罩　vacuum bell jar
真空轴密封　shaft seal
真实北　true north
真实气体临界流系数　real gas critical flow coefficient
振动保护　vibration protection
振动筛　vibrating screen/riddler/oscillating sieve
镇流器　ballast
蒸汽伴管　steam tracing
蒸汽流量　steam flow
蒸汽疏水阀　steam trap
整定电流　setting current
整定值　setting
整流器　rectifier
整体法兰　integral pipe flange
整体压缩机/摩托压缩机　integral compressor/motor compressor
正/负迁移　positive/negative elevation
正常时关闭　normally close
正常时开启　normally open
正火　normalizing
正极　positive pole

正交电压　quadrature voltage

正确度　trueness

正三通/Y 型三通　true “Y”

正视图　front view/facade

支撑点　point of support, POS

支承环　support ring

支承架　resting support

支耳　lug

支管连接　branch connection

支管台/凹台　boss

支架夹板　down lead

支架间距　support spacing

支腿　leg

支线　branch-line

执行机构　actuator

执行器　actuator

直柄立铣刀　vertical hobbing

直柄麻花钻　twist drill

直管　run pipe/straight pipe

直管段　straight length

直管螺纹　straight pipe thread

直角边缘孔板　square-edged orifice plate

直接雷击　direct lightning stroke

直径　diameter

直径比　diameter ratio

直流单臂电桥　DC single-arm electric bridge

直流电的正负极　DC positive and negative poles

直流电源　DC power supply

直流配电屏　DC distributing panel

直流配电线路　DC distribution circuit

直流稳压电源　Stabilized DC source

直流泄漏　DC leakage

直流主母线　DC main bus

直线度　straightness

止回阀　check valve

指示灯　indicator light

制动器　brake

制造厂(供货)　by vender

质量流量　mass flowrate

质量流量计　mass flowmeter

智能压力变送器　smart pressure transmitter

滞止压力　stagnation pressure

蛭石　vermiculite

中间继电器　auxiliary relay

中间接头　splice

中频电源　medium frequency power supply

中心至端面　center to end

中心至面　center to face

中心至中心　center to center

中心钻　center drill

中性点接地　neutral point grounded

中性线　neutral

中压　medium pressure

终点　ending

终端接头　end connector

钟罩式气体流量标准装置　standard bell prover

重量　weight

重瓦斯保护　heavy gas protection

重瓦斯预告信号　heavy gas prewarning signal

周长　circumference/perimeter

周缘流量　peripheral flow rate

轴测图　isometric drawing

轴流泵　axial flow pump

主回路　main circuit

主令开关　master switch/controller

主流向　mean direction of flow

主母线　main bus

主启动设备回路　circuit to starting device

主真空阀　main vacuum valve

贮藏室　storage

贮罐　tank

柱　column/post/stanchion

柱塞泵　plunger pump

柱塞阀　piston type valve

柱式管架　pole type support

铸钢　cast steel

铸件　casting

铸铁　cast iron

铸铁管　cast iron pipe

铸造　casting

专业　discipline

转鼓式过滤机　rotating drum type filter

转换开关　transfer switch

转换开关接点图　contact diagram of transfer switch

转换器　converter

转速表　tachometer

转子泵　rotary pump

转子流量计　rotameter/float meter

装配　assembly

装在吊钩上的插接式母线　hook-supported plug-in bus way

装在支架上的插接式母线　bracket-supported plug-in bus way

装在支柱上的插接式母线　post-supported plug-in bus way

装置北　plant north

装置边界　battery limit

锥柄锪钻　taper-shank spotfacer

锥柄扩孔钻　taper-shank fraise

锥柄立铣刀　taper-shank vertical hobbing

锥柄麻花钻　taper drill

锥度　conicity

锥管螺纹　taper pipe thread

锥形人口孔板　conical entrance orifice plate

准确度　accuracy

紫外线　ultraviolet

自保持　self-holding

自补偿旋桨　self-compensating propeller

自动操作继电器　relay for auto-operation

自动调度系统　automatic dispatch system，ACS

自动复位的操作开关　spring-return operating switch

自动跟踪　automatic track

自动化　automation

自动化仪表　automation instrumentation

自动开关　automatic switch

自动开关箱　automatic switch box
自动控制　automatic control
自动调整　automatic adjustment
自攻螺钉　self tapping screw
自控电压　power supply for process control
自耦变压器　auto-transformer
自适应控制 adaptive control
自锁　self-lock
自吸泵　self suction pump/self priming pump
自整角机　selsyn
综合控制系统　comprehensive control system
综合起动器　combination starter/magnetic starter combination
总流量　total flow
总平面布置图　general arrangement plan
总图　genera plan
总压皮托管　total pressure Pitot tube
总压取压孔　total pressure tapping
总压/滞压　stagnation pressure
阻火器　flame arrested/flame trap
阻尼　damping
阻尼计　snubber/damper
组合式热能表　combined heat meter
钻削　drilling
最大流量　maximum flow-rate
最小流量　minimum flow-rate
最终确认图纸资料　certified final
坐标　coordinate
坐标原点　origin of coordinate